The Calf

NUTRITION AND HEALTH

The Calf

NUTRITION AND HEALTH

J. H. B. Roy, M.A., Dip. Agric. (Cantab.), Ph. D.

The Pennsylvania State University Press
University Park and London

First Published 1955
Second Impression 1956
Second Edition 1959
Third Edition 1970

Published in the United States of America
by The Pennsylvania State University Press

Library of Congress Catalogue Card No. 71-119021
Standard Book No. 271-00122-4

Published in Great Britain by Iliffe Books,
an imprint of the Butterworth Group

Printed in Great Britain by
Cox & Wyman Ltd, Fakenham,
Reading and London

Contents

Preface

The field of coverage of this third edition of *The Calf–its Management, Feeding and Health* has been widened to provide sufficient background information for the specialist in calf nutrition and production. The first edition published in 1955 was kept as simple as possible and included only a small amount of the findings of agricultural science that were pertinent to an understanding of the methods of calf rearing then described. In enlarging the book, an effort has been made to maintain the advantages of the earlier editions, while increasing the value of the book for those who wish to have more detailed scientific information. To this end, the book has been divided into two volumes.

The first volume, *Management and Feeding*, contains much of the general information on feeding and management of calves that appeared in earlier editions, with appropriate revision of the text and inclusion of new material.

This, the second volume, *Nutrition and Health*, deals with the subject of nutrition in greater depth and also gives information on the more important calf diseases, especially those related to the nutritional and physical environment. Although it is impossible in a book of this size to cover all aspects of the nutrition of the calf, it is hoped that those readers who are interested in any particular aspect will find a sufficient range of references to scientific journals to satisfy their needs.

Shinfield, 1969 J. H. B. Roy

Acknowledgements

During the course of preparation of the earlier editions of this book, I was greatly assisted by a number of my colleagues at the National Institute for Research in Dairying, in particular Dr. K. W. G. Shillam and Mrs. G. Maddox.

The preparation of this third edition has been considerably eased by the valuable help that I have received from a number of assistants in the conduct of experiments with calves over recent years. I wish also, to thank Mrs. P.D.A. McMillan Browse for her loyal services, Dr. I.J.F. Stobo for his advice and for checking the manuscript, Mrs. E. B. Poulter, who has managed to translate illegible handwriting into typescript, and Mr. B. F. Bone for assisting in the checking of the large number of references.

PART 1

Nutrition

CHAPTER ONE

The Digestive System of the Calf

THE PRE-RUMINANT CALF

The young calf at birth, like the adult ruminant, has a stomach of four parts, although only the abomasum or fourth stomach, with a capacity about twice that of the other compartments, is functional.[1] The reticulo-rumen, although non-functional, has an inherent capacity of about 2 l. (3·5 pt) at birth.[2] In the adult ruminant, however, only about 8% of the total capacity is in the abomasum, whereas the volume of the rumen represents 80% of the total.[3]

In the young calf, liquid food can bypass the reticulo-rumen and go straight to the abomasum, by flowing through a tube formed by the closure of the oesophageal groove (see Fig. 1.1). The reflex closure of this groove, which extends from the cardia to the reticulo-omasal orifice, occurs when the soluble proteins and salts of milk are given,[4] and can be brought about by stimulation of the glossopharyngeal nerve.[5] Of salt solutions tested, 60 ml of a 10% solution of sodium bicarbonate was the most effective in causing closure.[6] With some animals, glucose invariably passed into the abomasum. It has been suggested that the reflex closure of this groove is probably evoked as part of the sucking pattern of behaviour, and not as part of the thirst pattern, for water does not elicit as strong an effect as milk.[7] A great variation, however, exists between calves in the response to various stimuli and the reflex action becomes weaker as the calf becomes older. Until the calf is about 8 weeks old, the oesophageal groove reflex is stimulated by water, as well as by milk and milk by-products, but after this time water becomes less effective.

Up to the age of 12 weeks, milk is effective whether given by nipple feeder or open bucket, but after this age the nipple feeder is superior. Capsules given to calves between liquid meals pass into the rumen, whereas those given at the same time as the calf is drinking pass into the abomasum.[8]

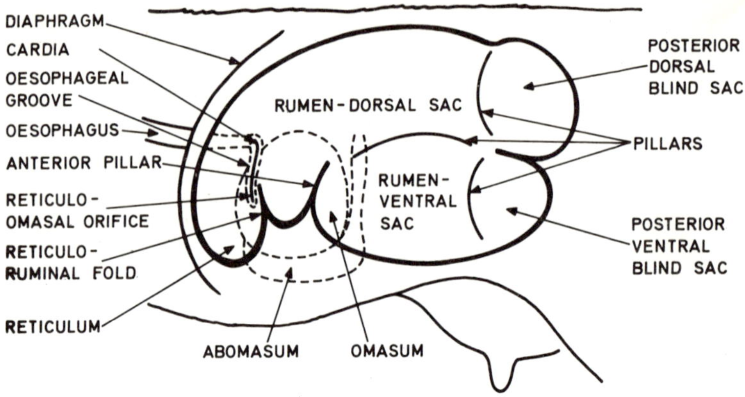

Fig. 1.1. Diagrammatical section through the rumen and reticulum of a cow showing the approximate position of the omasum and abomasum[101]

During the first 4 weeks of life, it seems that the only nutrients that can be satisfactorily utilised when given in liquid form are milk proteins, butterfat or vegetable and other animal fats, and the sugars, lactose (milk sugar) and glucose (see p. 6). In veal production, the calf may be made dependent on the products of monogastric digestion throughout its life, by withholding dry food.

ENZYME ACTIVITY IN SALIVA

It has been shown that a lipase is produced in the saliva of the young calf, and this lipase only acts on the triglycerides of butterfat that contain butyrate groups to release butyric acid. This lipase has an optimum pH of 4·5–6·0 and its activity decreases with increasing age of the calf and normally disappears by the third month, the decrease being more rapid with calves given high-roughage diets. Its production is stimulated by a calf either sucking or drinking milk, although a slow intake of milk such as occurs by sucking through a nipple appears to cause greater stimulation.[9–12]

ENZYME ACTIVITY IN THE ABOMASUM AND CLOTTING OF MILK PROTEIN

After milk is ingested by the calf, clotting takes place within 1–10 min[13] as a result of the enzyme action of either rennin or pepsin. During the first 3–4 hr after feeding whey is released from the clot into the abomasum and passes into the duodenum, followed thereafter by partially digested casein.[14] Impairment of rennin clotting seems to be an important factor in predisposing calves to an *Escherichia coli* localised intestinal infection (see p. 123).

Originally it was thought that the milk-fed calf secretes only rennin and that the secretion of pepsin does not become marked until the calf eats solid food,[15,16] or at least until 4 weeks of age.[17] However, further studies[18] have suggested that the young calf may secrete either rennin or pepsin, or both, and that the exact pattern is not predictable from the age of the animal or the nature of its diet.

The pH of the abomasum before feeding is about 2–2·8, but rises rapidly to reach values of 4·5–6·2, 30 min after feeding milk and then declines to pre-feeding levels at 3–5 hr.[14] However, it appears that age affects pH, a value of 4·1 in new-born calves 4 hr after feeding had fallen to 3·4 at 16 days of age.[19] In the new-born calf, it has been suggested that the pH of the abomasal contents, after sucking does not reach a level sufficiently low for effective proteolysis.[20] From *in vitro* studies it is clear that for clotting by acid alone to occur, the concentration of acid needed is twice that required in the presence of pepsin. It seems probable therefore that clotting of milk at a very young age is not caused by the pepsin–HCl system. Moreover, pasteurised milk elicits less secretion of HCl into the abomasum of calves than does fresh milk.[21] The output of acid and proteolytic activity in the abomasum, however, increases as the quantity of milk given to a calf increases, but it tends to lag behind the increase in intake.[22] Thus the effect of large volumes of whole milk in predisposing some calves to diarrhoea at a young age may well be the result of swamping a limited enzyme production.

SUBSTITUTION OF MILK BY VEGETABLE PROTEIN

The substitution of expensive milk protein by vegetable protein in the diet of pre-ruminant calves has not been very successful. Most experiments have been done with soya bean protein, and in general the cruder the product, the worse have been the results.[23] Ground, cooked soya beans in place of skim milk gave little weight gains in a 100-day period but purified soya protein gave much better results.[24] Recently it has been claimed[25,26] that a 71% protein soya flour providing 86% of the protein of a milk substitute gave as good results as whole milk, but the level of feeding was such that the Friesian calves given whole milk gained weight at 0·20 kg/day (0·44 lb/day) and the soya bean-fed calves at 0·33 kg/day (0·74 lb/day). It was suggested that the good performance on this soya diet was due to its low content of trypsin inhibitor, which is normally present in the crude product. When a soya flour containing 50% protein was used, growth was negligible, due either to diarrhoea or to the presence of a trypsin inhibitor. This poor performance on the 50% protein soya diet was reflected in a marked decrease in the volume of secretion, and protein and enzyme concentration of the pancreatic juice in the calves compared with that of calves given skim milk. More recently it has been claimed that acid or alkali treatment of soya flour markedly improves its utilisation by the calf.[27, 28] Milk substitute diets containing these treated soya flours as the sole source of protein have resulted in growth rates of up to 0·57 kg/day (1·25 lb/day) in the first 8 weeks of life.

ENZYME ACTIVITY IN THE INTESTINE

As mentioned above the only carbohydrates of use to the very young calf are glucose and lactose.

It appears that the pre-ruminant calf cannot digest starch or the degradation products, dextrin and maltose, up to 28 days of age, as the enzymes amylase and maltase are only present in the pancreas and intestine, respectively, in very low concentration, although they begin to increase by 6–9 weeks of age.[29–31]

No evidence has been shown that calves reared on starch-containing diets from a very young age show any increased enzyme adaptation.[32,33] Diets containing starch may cause diarrhoea and, moreover, a condition has been reported in young calves given starch, which results in the animals being depressed and refusing their food. The condition is associated with the fermentation of starch in the gut and the presence of large numbers of yeast cells and a high blood alcohol concentration.[34] On the other hand, cereal flours may on occasions have some benefit other than as actual nutrients, for cornflour and arrowroot have for many years been used as remedies for diarrhoea.

The possibility of feeding enzymes to pre-ruminant calves to improve their utilisation of starch has been studied and has shown some success when the starch is fully gelatinised. With a milk substitute containing ground sorghum grain that had been expanded and in which such gelatinisation had occurred, the addition of the enzyme amyloglucosidase gave a good response in blood glucose level after feeding. Moreover, calves given this diet gained weight during the first 7 weeks of life at the same rate as those given a milk substitute containing glucose, although the level of feeding was such as only to give a weight gain of 0·54–0·58 kg/day (1·2–1·3 lb/day).[35]

Sucrose (cane sugar) is very poorly utilised by the calf[30,36] as no intestinal sucrase activity has been observed.[29] The small amounts of sucrose that are absorbed from the intestine are probably the result of degradation by intestinal micro-organisms,[37] which are also probably responsible for the apparent digestion of a small amount of starch[38] in the young calf even in the absence of enzymes.

Lactase activity in the intestine declines with age,[31] but in this case, adaptation to diet can occur. By rearing calves on whole milk supplemented with 15% lactose, a three-fold increase in intestinal lactase was observed over that in unsupplemented calves.[39] Even in ruminant calves. in which the rumen has been bypassed, lactose has been found to be much better utilised than starch in the intestine.[33]

By feeding meal mixtures as a gruel some of the advantages of monogastric digestion can be achieved. A much higher retention of digested nitrogen was obtained in 14-week old calves when a concentrate mixture was given as a gruel rather than in dry form,[40] a finding probably due to the more efficient use of dietary energy in monogastric digestion (see p. 20).

THE RUMINANT CALF

RUMEN DEVELOPMENT

The age at which transition to the ruminant method of digestion occurs is largely dependent on the diet that a calf receives. The longer the period that a calf has access to a plentiful supply of milk, the less will be its urge to supplement its diet with other foods. Indeed, veal calves given *ad lib.* liquid diet and access to dry food consumed only about 3 kg (6·6 lb) hay in a 3 month period.[41] Given a limited quantity of liquid diet, calves will begin eating grass and, providing it is palatable, dry food at about 7 days of age. These dry foods pass into the rumen, where bacteria and other micro-organisms become established and convert fibrous and starchy foods into an available form of energy, mainly as volatile fatty acids (VFA), and synthesise B vitamins and create protein from simpler nitrogenous compounds. Once the rumen begins to function, the incidence of diarrhoea of calves given a balanced dry ration is negligible.

Calves having a restricted concentrate intake but access to large amounts of roughage have a considerably increased capacity of the reticulo-rumen, but this increase has been found to be largely the result of stretching the tissues.[42] However, the inclusion of dry food, besides increasing the capacity of the reticulo-rumen also increases the weight of tissue.[2, 43] Only a relatively small increase occurs in the thickness of the muscular wall, but there is a considerable increase in thickness of the mucosa, due to the development of rumen papillae ('finger-like' projections), which is much greater in concentrate-fed calves than in those given large amounts of hay and other roughages (see Plate 1.1).[44] The papillae increase the surface area of the rumen walls and thus the area through which nutrients can be absorbed. Their development is stimulated by the end-products of rumen fermentation rather than by the fibrous nature of the food.[45-47] Solutions of sodium butyrate and, to a less extent, sodium propionate cause extensive papillary growth, whereas sodium acetate has decidedly less effect.[48,49] Even so, calves normally start eating their fibrous bedding before they begin eating concentrates.

Excessive papillary development may lead to rumen parakeratosis.[50-55] In this condition, there is an increase in the length and width of the papillae, clumping of the papillae with rumen contents and encrusting of the tips of the papillae with dark keratinised material. The condition occurs with high-concentrate diets, or mixed diets in which the hay has been pelleted, rather than wafered or given in long form. The low rumen pH on such diets, together with an increase in production of propionic acid relative to acetic acid, and lack of muscle tone have all been associated with the condition. Absorption of volatile fatty acids through the rumen wall may be reduced when the condition becomes severe.

The extent of the papillary development within the normal range has no demonstrable effect on the calf in its ability to digest diets of concentrates and hay. A period as short as 3 weeks is sufficient for increased papillary development to occur in calves transferred from a high-roughage diet to one of all concentrates, and for retrogression of papillae if transferred from a high-concentrate diet to one of all hay.[56] Comparison of high- or low-roughage diets in the rearing period from 4 months to 2 years of age, showed no effect of the method of rearing on the ability of the heifers to digest hay or mixed rations of hay and concentrates.[57] On the other hand, experience indicates that heifers reared on very high levels of high-concentrate diets may be loath to eat hay when subsequently given the opportunity to do so.

The nervous mechanism controlling the various muscular phenomena of rumination is very active in the 2-week old calf and only slight stimulation, such as the presence of recording balloons, is necessary to produce strong and frequent regurgitation effects.[58] Practically normal rumen contractions in a calf as early as 4 days of age have been demonstrated.[59] When given limited quantities of liquid diet, some calves may ruminate by 5 days of age, and the majority do so by 28 days of age. The time spent ruminating increases rapidly in calves given hay and concentrates to average 5 hr/day at 6–8 weeks of age.[60]

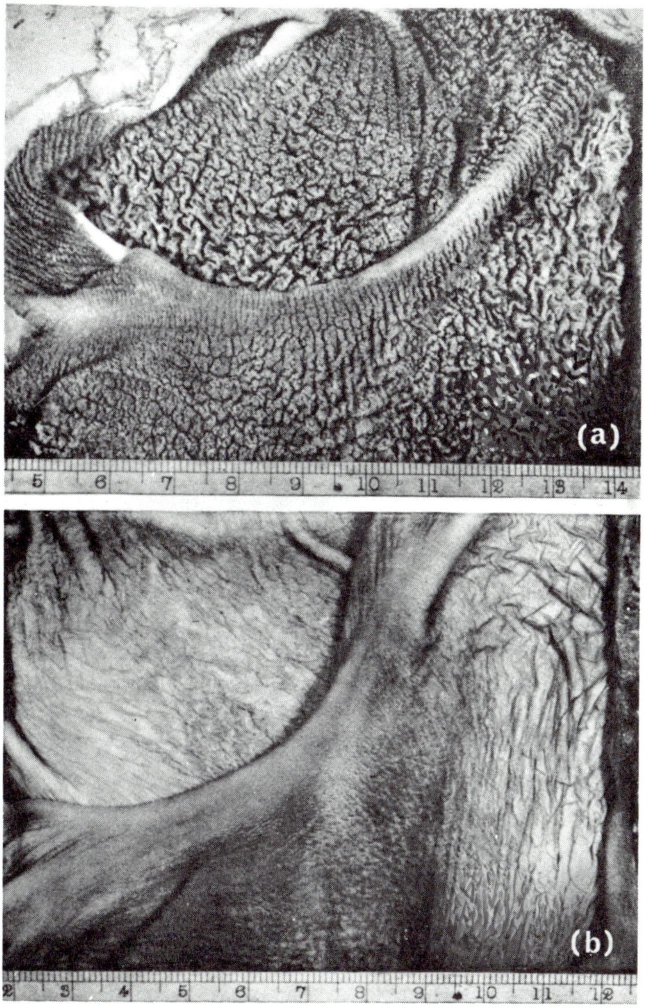

Plate 1.1. Effect of concentrates and hay on papillary development of the rumen. Papillae development in the ventral blind sac of the reticulo-rumen of two 12-week old calves. (a) Calf had eaten 30 kg (66 lb) barley and 7 kg (15 lb) of hay; (b) calf had eaten 0·2 kg (0·5 lb) of barley and 25 kg (55 lb) of hay. Both calves were also given liquid skim milk ad lib.

GUT FILL

The mode of digestion, whether monogastric or ruminant, has a marked effect on the weight of contents of the alimentary tract and thus on the relationship between live weight and empty body weight. With pre-ruminant calves, the total weight of contents has been shown to be 4·1%, 8 hr after the last feed at 7 days of age;[19] 6·2%, 3 hr after the last feed at 21 days of age;[43] and 3·6%, 17 hr after the last feed, for veal calves at 104 days of age.[61] However, on *ad lib.* feeding of liquid diets, these values may be increased if the dry matter content of the diet is low, and thus a greater volume is consumed.

On the other hand with ruminant calves at 12 weeks of age, the values are very much higher, varying from 27% on a diet of *ad lib.* hay to 11% on a diet of *ad lib.* concentrates. An attempt to predict[43] the weight of the contents of the alimentary tract in the live animal gave the following equation:

$$y = 0{\cdot}20x_1 - 3{\cdot}33x_2 + 0{\cdot}40x_3 + 2{\cdot}90$$

where y is the weight of contents (kg), x_1 the live weight (kg), x_2 the daily concentrate intake (kg), and x_3 the daily hay intake (kg).

In addition, the weight of the alimentary tract itself is much greater in the ruminant calf, the total weight of contents and alimentary tract being as high as 36% of live weight for all-roughage diets and 19% for all-concentrate diets, compared with 11% for the pre-ruminant calf.

The effect of gut fill should be considered when comparing the performance of calves and economics of calf rearing methods in which very different types of rations are being used.

DEVELOPMENT OF THE RUMEN FLORA

The abomasum of the very young calf supports a large and diverse population of lactobacilli[62] (see p. 120) and it is possible that a backflow of milk from the abomasum to the rumen after feeding[63–65] may help to inoculate the rumen.

The typical rumen amylolytic streptococci of the adult animal do not become properly established until the rumen pH is stabilised near neutrality and under conventional rearing methods, this does not happen until the calf is several weeks old, although in the very young calf a different amylolytic streptococcus is found which is more tolerant of a lower pH value.[62] Similarly, the protozoan population can become established once the pH is stabilised near neutrality at about 8 weeks of age.[66]

When the proportion of concentrates to hay in a ration exceeds a ratio of about 3:1, it has been shown that the types of bacteria associated with the digestion of hay disappear.[67] Whereas protozoa are encouraged by high-roughage diets, lactobacilli tend to be encouraged by high-concentrate diets.[68] When calves were changed from a high-concentrate diet to one of all hay, a marked reduction in the viable count of streptococci, lactobacilli and *E. coli* was found, but when the change was from a high-roughage to an all-concentrate diet no effect on the bacterial flora was apparent.[56]

RUMEN INOCULATION

To ensure a normal microbial population in the calf, it has been suggested that there is a need for rumen inoculation with the cud of an adult animal. In Israel[69] and in Russia[70] beneficial effects of inoculation on growth rate have been claimed but later experiments have not confirmed this.[71–81]

However, it has been shown[82] that inoculation can hasten the development of protozoa in the rumen, although the development of rumen bacteria is not affected unless the calf is raised under strictly isolated conditions. In general, it seems that, under normal farm conditions, where roughage is made freely available to the calf in its early life, or where the calf has access to pasture, no benefit will accrue from cud inoculation, the calf developing its normal bacterial flora without assistance. With calves eating a high proportion of concentrates in relation to hay, so that the growth of protozoa are inhibited, inoculation with protozoa will not increase their numbers.[67]

DEVELOPMENT OF DIGESTION IN THE RUMEN

DIGESTIBILITY

Concentrates or good quality grass seem to be as well digested by the young calf as by the adult ruminant[77,83–85] but there appears to be an increase with age in the digestibility of hay or high-roughage diets.[86,87] At an age of 13–17 weeks no associative effects were recorded between the digestibility of hay and concentrates when fed separately or in combination.[56]

VOLATILE FATTY ACID PRODUCTION

The concentration of total VFA produced by rumen fermentation and supplying a major source of energy to the ruminant, increases with the age of the calf, the time of the maximum value occurring about a week after weaning.[86, 88–91] However, no difference in concentration of total

VFA's has been found in the rumen between 8 and 14 weeks, but as there was certainly an increase in volume of rumen contents, the total production of VFA must have increased.[56] There is some evidence that pre-ruminant calves with a non-functional rumen are able to utilise VFA's, which are probably produced in the large intestine.[92]

RUMEN pH

The pH of the rumen contents has been shown to be low at weaning, with calves weaned at 3 weeks of age, but thereafter the pH increases, with age.[91] The low pH may be due to an increased production of lactic acid at a young age or to a low level of saliva production.[93] The greatest concentration of lactic acid in the rumen occurs at about 1 month of age[88] and a slightly higher proportion was found in calves given all-concentrate diets compared with those given an all-hay diet at 16 weeks.[56]

CONCENTRATION OF INDIVIDUAL VOLATILE FATTY ACIDS IN THE RUMEN

In the developing rumen, the proportion of acetic acid has been found to be lower than in adult cows[86] and did not differ between calves given hay or concentrates; the proportion of butyric acid increased rapidly from 10% at 4 weeks to 20–30% thereafter. The changes in proportion of acids as a result of different diets tend to be variable. For instance, values of 72–64% for acetic acid have been obtained, with rations varying in hay concentrate ratio from 4 : 1 to 2 : 3, respectively,[94] and as the amount of concentrates in the diet increased, butyric acid increased at the expense of propionic acid. However, in another experiment, the proportion of propionic to acetic acid in the rumen was very much higher in calves given high-concentrate diets than those given high-roughage diets, and the proportion of butyric acid was much less affected by diet.[56]

The possibility of using salts of VFA as additives to diets has been explored and slight increases in growth rate have been reported[95] but their commercial value for calves is doubtful, especially in view of their effect on rumen parakeratosis (see above).

BLOOD VFA LEVELS

The VFA concentration in peripheral blood increases with age,[89] but tends to be higher in calves given high-roughage or hay diets than in those given high-concentrate or all-concentrate diets.[56]

BLOOD GLUCOSE LEVELS

In the young calf, blood glucose levels fall from a value of about 100 mg/100 ml to about 60 mg/100 ml at 6 weeks of age.[96,97] It has been suggested[98] that this decline is associated with the replacement of foetal erythrocytes (red cells) high in glucose with erythrocytes of lower glucose content, or possibly also by an increase in the production of VFA in the large intestine,[92] on the basis that the administration of propionate may cause a fall in blood glucose.[99]

In the pre-ruminant calf, blood glucose rises rapidly after a meal to a peak of about 130 mg/100 ml at 1–2 hr after feeding, and then declines to pre-feeding values about 5 hr after feeding.[29]

In the ruminant calf, however, blood glucose levels are much more stable, a slight depression occurs about 3 hr after feeding followed by a gradual rise to maximum values 8–12 hr after feeding.[100] Much higher levels are, however, maintained by calves given all-concentrate diets than by those given all-hay diets. At 7 weeks of age, milk-fed and concentrate-fed calves had mean blood glucose levels of 75 mg/100 ml, whereas hay-fed calves declined to 55 mg/100 ml at 8 weeks.[2,45] At 16 weeks of age, calves on all-concentrate diets had mean values of 93 mg/100 ml compared with 68 mg/100 ml for calves on all-hay diets.[56]

How far these high plasma glucose levels of calves on concentrates as

Table 1.1 COMPARISON OF CERTAIN CHARACTERISTICS OF 16-WEEK OLD AYRSHIRE CALVES GIVEN CONCENTRATES OR HAY *AD LIB.* (1 kg = 2·2 lb) (BASED ON REFERENCE 56)

	Diet	
	Concentrates ad lib.	*Hay ad lib.*
Dry matter intake (kg/day)	2·5	1·6
Digestibility of dry matter (%)	81	58
Rumen total volatile fatty acids (m-equiv./100 ml)	11	5
Partition of rumen VFA (molar %)		
Acetic	47	74
Propionic	31	19
Butyric	18	7
Valeric and higher	4	–
Rumen pH	5·9	7·4
Plasma total VFA (mg/100 ml)	0·9	1·3
Blood glucose (mg/100 ml)	93	56
N-retention (g/day)	31	4
Live weight gain (kg/day)	0·65	0·19

opposed to hay diets are the result of the glucogenic effect of propionate, or to glucose formed by hydrolysis of starch in the intestine is uncertain, as little is known of the extent to which concentrates can escape digestion in the rumen by rapid passage into the abomasum. A comparison of certain characteristics of 16-week old calves given concentrates or hay *ad lib.* are shown in Table 1.1.

REFERENCES

1. GROSSMAN, J. D. *J. Am. vet. med. Ass.* **114**, 416 (1949)
2. WARNER, R. G., FLATT, W. P. and LOOSLI, J. K. *J. agric. Fd Chem.* **4**, 788 (1956)
3. SISSON, S. *The Anatomy of the Domestic Animals* 4th edn (W. B. Saunders Co., 1953)
4. WESTER, J. *Vet. J.* **86**, 401 (1930).
5. COMLINE, R. S. and TITCHEN, D. A. *J. Physiol., Lond.* **115**, 210 (1951)
6. RIEK, R. F. *Aust. vet. J.* **30**, 29 (1954)
7. WATSON, R. H. *Bull. Coun. scient. ind. Res., Melb.* **180**, 1 (1944)
8. HEGLAND, R. B., LAMBERT, M. R., JACOBSON, N. L. and PAYNE, L. C. *J. Dairy Sci.* **40**, 1107 (1957)
9. GROSSKOPF, J. F. W. *Onderstepoort J. vet. Res.* **32**, 153 (1965)
10. YOUNG, J. W., RAMSEY, H. A. and WISE, G. H. *J. Dairy Sci.* **43**, 1068 (1960)
11. RAMSEY, H. A. *J. Dairy Sci.* **45**, 1479 (1962)
12. OTTERBY, D. E., RAMSEY, H. A. and WISE, G. H. *J. Dairy Sci.* **47**, 993 (1964)
13. MORTENSON, F. N., ESPE, D. L. and CANNON, C. Y. *J. Dairy Sci.* **18**, 229 (1935)
14. MYLREA, P. J., *Res. vet. Sci.* **7**, 333 (1966)
15. BERRIDGE, N. J., DAVIS, J. G., KON, P. M., KON, S. K. and SPRATLING, F. R. *J. Dairy Res.* **13**, 145 (1942–44)
16. DOLLAR, A. M. 'Biochemical studies of nutrition in the ruminant'. *Ph.D. Thesis.* (University of Reading, 1958)
17. HENSCHEL, M. J., HILL, W. B. and PORTER, J. W. G. *Proc. Nutr. Soc.* **20**, xl (1961)
18. PORTER, J. W. G. Personal communication (1963)
19. KESLER, E. M., RONNING, M. and KNODT, C. B. *J. Anim. Sci.* **10**, 969 (1951)
20. PIERCE, A. E., *Colston Pap.* **13**, 189 (1962)
21. JANCARIK, A., *Sb. čsl. Akad. zeměd. Věd Zivoč. V́ yr.* **7**, 71 (1962)
22. ASH, R. W., *J. Physiol., Lond.* **172**, 425 (1964)
23. PORTER, J. W. G., *Rep. natn. Inst. Res. Dairy* p.126 (1963)
24. BENEVENGA, N. J. and RONNING, M. *J. Anim. Sci.* **22**, 832 (1963)
25. GORRILL, A. D. L., THOMAS, J. W., STEWART, W. E. and MORRILL, J. L. *J. Nutr.* **92**, 86 (1967)
26. GORRILL, A. D. L. and THOMAS, J. W. *J. Nutr.* **92**, 215 (1967)
27. COLVIN, B. M. and RAMSEY, H. A. *J. Dairy Sci.* **51**, 898 (1968)
28. COLVIN, B. M. and RAMSEY, H. A. *J. Dairy Sci.* **52**, 270 (1969)
29. DOLLAR, A. M. and PORTER, J. W. G. *Nature, Lond.* **179**, 1299 (1957)
30. OKAMOTO, M., THOMAS, J. W. and JOHNSON, T. L. *J. Dairy Sci.* **42**, 920 (1959)

31. DOLLAR, A. M. and PORTER, J. W. G. *Int. Dairy Congr. XV,* London, **1**, 185 (1959)
32. HUBER, J. T., JACOBSON, N. L., ALLEN, R. S. and HARTMAN, P. A. *J. Dairy Sci.* **44**, 1494 (1961)
33. HUBER, J. T., NATRAJAN, S. and POLAN, C. E. *J. Dairy Sci.* **50**, 1161 (1967)
34. ABE, R. K., MORRILL, J. L., BASSETTE, R., MUSSMAN, H. C. and OEHME, F. W. *J. Dairy Sci.* **49**, 727 (1966)
35. MORRILL, J. L., ABE, R. K. and DEYOE, C. W. *J. Dairy Sci.* **49**, 727 (1966)
36. VELU, J. G., GARDNER, K. E. and KENDALL, K. A. *J. Dairy Sci.* **42**, 920 (1959)
37. XENOULIS, P. K., JACOBSON, N. L., YOUNG, J. W., McGILLIARD, A. D. and ALLEN, R. S. *J. Anim. Sci.* **26**, 933 (1967)
38. SHAW, R. H., WOODWARD, T. E. and NORTON, R. P. *J. agric. Res.* **12**, 575 (1918)
39. HUBER, J. T., RIFKIN, R. J. and KEITH, J. M. *J. Dairy Sci.* **47**, 789 (1964)
40. RAVEN, A. M. and ROBINSON, K. L., *Nature, Lond.* **192**, 1256 (1961)
41. ROY, J. H. B., STOBO, I. J. F. and GASTON, H. J. *Rep. natn. Inst. Res. Dairy.* p. 53 (1966)
42. BLAXTER, K. L., HUTCHESON, M. K., ROBERTSON, J. M. and WILSON, A. L. *Br. J. Nutr.* **6**, i (1952).
43. STOBO, I. J. F., ROY, J. H. B. and GASTON, H. J. *Br. J. Nutr.* **20**, 171 (1966)
44. BROWNLEE, A. *Br. vet. J.* **112**, 369 (1956)
45. WARNER, R. G., BERNHOLDT, H. F., GRIPPIN, C. H. and LOOSLI, J. K. *J. Dairy Sci.* **36**, 599 (1953)
46. WARNER, R. G., GRIPPIN, C. H., FLATT, W. P. and LOOSLI, J. K. *J. Dairy Sci.* **38**, 605 (1955)
47. FLATT, W. P., WARNER, R. G. and LOOSLI, J. K. *J. Dairy Sci.* **41**, 1593 (1958)
48. SANDER, E. G., WARNER, R. G., HARRISON, H. N. and LOOSLI, J. K. *J. Dairy Sci.* **42**, 1600 (1959)
49. TAMATE, H., McGILLIARD, A. D., JACOBSON, N. L. and GETTY, R. *J. Dairy Sci.* **45**, 408 (1962)
50. VIDACS, G. and WARD, G. M. *J. Dairy Sci.* **43**, 875 (1960)
51. GARRETT, W. N., MEYER, J. H., LOFGREEN, G. P. and DOBIE, J. B. *J. Anim. Sci.* **20**, 833 (1961)
52. WARD, G. M. *Feedstuffs, Minneap.* **34**, i (1962)
53. GILLILAND, R. L., BUSH, L. J. and FRIEND, J. D. *J. Dairy Sci.* **45**, 1211 (1962)
54. BULL, L. S., BUSH, L. J., FRIEND, J. D., HARRIS, B. Jr. and JONES, E. W., *J. Dairy Sci.* **48**, 1459 (1965).
55. GIRARDI, V. and PATRIZI, R. *Atti Soc. ital. Sci. vet.* **19**, 325 (1966)
56. STOBO, I. J. F., ROY, J. H. B. and GASTON, H. J. *Br. J. Nutr.* **20**, 189 (1966)
57. BALCH, C. C., CAMPLING, R. C., JOHNSON, V. W. and ROY, J. *Br. J. Nutr.* **14**, 379 (1960)
58. SCHALK, A. F. and AMADON, R. S. *Bull. N. Dak. agric. Exp. Stn* no. 216 (1928)
59. ROY, J. H. B., *Outl. Agric.* **2**, 219 (1959)
60. SWANSON, E. W. and HARRIS, J. D. *J. Dairy Sci.* **41**, 1768 (1958)
61. ROY, J. H. B., GASTON, H. J., SHILLAM, K. W. G., THOMPSON, S. Y., STOBO, I. J. F. and GREATOREX, J. C. *Br. J. Nutr.* **18**, 467 (1964)
62. MANN, S. O. and OXFORD, A. E. *J. gen Microbiol.* **12**, 140 (1955)

63. TRAUTMANN, A. and SCHMITT, J. *Arch. Tierernähr. Tierz.* **9**, 11 (1933)
64. HOFLUND, S. *Svensk VetTidskr.* **51**, 88 (1946)
65. BENZIE, D. and PHILLIPSON, A. T. *The Alimentary Tract of the Ruminant.* (Oliver and Boyd, Edinburgh, 1957)
66. MANN, S. O., MASSON, F. M. and OXFORD, A. E. *Br. J. Nutr.* **8**, 246 (1954)
67. POUNDEN, W. D. and HIBBS, J. W. *J. Dairy Sci.* **31**, 1051 (1948)
68. EADIE, J. M., HOBSON, P. N. and MANN, S. O. *Nature, Lond.* **183**, 624 (1959)
69. VOLCANI, R. and LEVI, D. *Refuah vet.* **10**, 54 (1953)
70. BELENKI, N. G. *Veterinariya* **30**, 25 (1953)
71. PELISSIER, C. L., SLACK, S. T., TRIMBERGER, G. W., TURK, K. L. and LOOSLI, J. K. *Fm Res.* **20**, no. 2 (1954)
72. DAVIS, L. R., AUTREY, K. M., HERLICH, H. and HAWKINS, G. E. J. *J. Dairy Sci.* **37**, 562 (1954)
73. GRIMBERGEN, A. H. M. *Meded. Inst. Moderne Veevoeding 'De Schothorst',* **S 42** (1955)
74. WILLIAMS, J. B. and JENSEN, C. *Bi-m Bull. N. Dak. agric. Exp. Stn* **17**, 91 (1955)
75. WIERINGA, G. W. *Tijdschr. Diergeneesk.* **81**, 242 (1956)
76. HARDISON, W. A., MILLER, G. A. and GRAF, G. C. *J. Dairy Sci.* **40**, 363 (1957)
77. McARTHUR, A. T. G. *N.Z. Jl Sci. Technol.* **38A**, 696, 700 (1957)
78. RONNING, M., BANKSTON, B. G. and BEROUSEK, E. R. *Tech. Bull. Okla. agric. Exp. Stn* no. T68 (1957)
79. BRÜGGEMANN, H. *Züchtungskunde* **30**, 179 (1958)
80. HIBBS, J. W. and CONRAD, H. R. *J. Dairy Sci.* **41**, 1230 (1958)
81. PRESTON, T. R. *Proc. Br. Soc. Anim. Prod.* p. 33 (1958)
82. BRYANT, M. P. and SMALL, N. *J. Dairy Sci.* **39**, 927 (1956)
83. ARMSTRONG, D. G., PRESTON, T. R. and ARMSTRONG, R. H. *Nature, Lond.* **174**, 1182 (1954)
84. HOGUE, D. E., WARNER, R. G., GRIPPIN, C. H. and LOOSLI, J. K. *J. Anim. Sci.* **15**, 788 (1956)
85. PRESTON, T. R., ARCHIBALD, J. D. H. and TINKLER, W. *J. agric. Sci., Camb.* **48**, 259 (1957)
86. FLATT, W. P., WARNER, R. G. and LOOSLI, J. K. *Mem. Cornell Univ. agric. Exp. Stn* no. 361 (1959)
87. MÄKELÄ, A. *Maataloust. Aikakausk.* **32**, 161 (1960)
88. LENGEMANN, F. W. and ALLEN, N. N. *J. Dairy Sci.* **38**, 651 (1955)
89. McCARTHY, R. D. and KESLER, E. M. *J. Dairy Sci.* **39**, 1280 (1956)
90. HIBBS, J. W., CONRAD, H. R., POUNDEN, W. D. and FRANK, H. *J. Dairy Sci.* **39**, 171 (1956)
91. DINDA, P. K. 'Some effects of chlortetracycline on the nutrition of the early weaned calf' *Ph.D Thesis.* (University of Aberdeen 1960)
92. LIANG, Y. T., MORRILL, J. L. and NOORDSY, J. L. *J. Dairy Sci.* **50**, 1153 (1967)
93. PRESTON, T. R. *Wld Rev. Nutr. Diet.* **4**, 117 (1963)
94. HIBBS, J. W., CONRAD, H. R. and POUNDEN, W. D. *J. Dairy Sci.* **37**, 724 (1954)
95. MARTIN, W. G., RAMSEY, H. A., MATRONE, G. and WISE, G. H. *J. Dairy Sci.* **42**, 1377 (1959)
96. McCANDLESS, E. L. and DYE, J. A. *Am. J. Physiol.* **162**, 434 (1950)
97. DYE, J. A. and ORSINI, D. *Fedn Proc. Fedn Am. Socs exp. Biol.* **11**, 39 (1952)

98. VANDERSALL, J. H., CONRAD, H. R. and HIBBS, J. W. *J. Anim. Sci.* **16,** 1111 (1957)
99. YOUNG, J. W., TOVE, S. B. and RAMSEY, H. A. *J. Dairy Sci.* **48,** 1079 (1965)
100. PRESTON, T. R. and NDUMBE, R. D. *Br. J. Nutr.* **15,** 281 (1961)
101. BALCH, C. C., *Outl. Agric.* **2,** 33 (1958)

CHAPTER TWO

The Nutrient Requirements of the Calf — Appetite, Water, Energy and Protein

APPETITE

The maximum voluntary dry matter intake of palatable foods by the young calf is shown in Fig. 2.1 and is dependent on the form in which the food is given. Up to a live weight of 70 kg (154 lb), more dry matter can be consumed in liquid than in dry form. Moreover, because the gross energy of liquid diets is utilised more efficiently by monogastric digestion than that of dry diets by ruminant digestion, the same amount of dry matter yields more net energy when given in liquid form. Thus, up to about 100 kg (220 lb) empty body weight, the calf will be able to obtain more net energy when the diet is in the form of milk, even though the intake at this weight is only 2·0 kg (4·4 lb) dry matter for the pre-ruminant calf, but 3·3 kg (7·3 lb) for the ruminant calf.

The maximum dry matter intake increases from about 2·2 kg/100 kg live weight for the 50 kg (110 lb) calf to 3·0 kg/100 kg for the 120 kg (264 lb) calf, the latter value being the generally accepted maximum dry matter intake for adult ruminants.

The consumption of liquid diets is, however, markedly influenced by their dry matter content. With diets of low dry matter content, such as liquid skim milk, a greater volume will be consumed than with whole milk, but the total dry matter intake will be very similar. Calves of a particular weight tend to consume dry matter to achieve a particular calorie intake, but may be prevented from doing this if the volume that has to be consumed is too great. Thus, when a 10% fat milk substitute was reconstituted at levels of 5–25% dry matter, the greatest volume was consumed at the lowest reconstitution rate, but the calves achieved the highest dry matter intake at a reconstitution rate of 15% and above.[1] Although the reconstitution rate of 15% was found to give the greatest

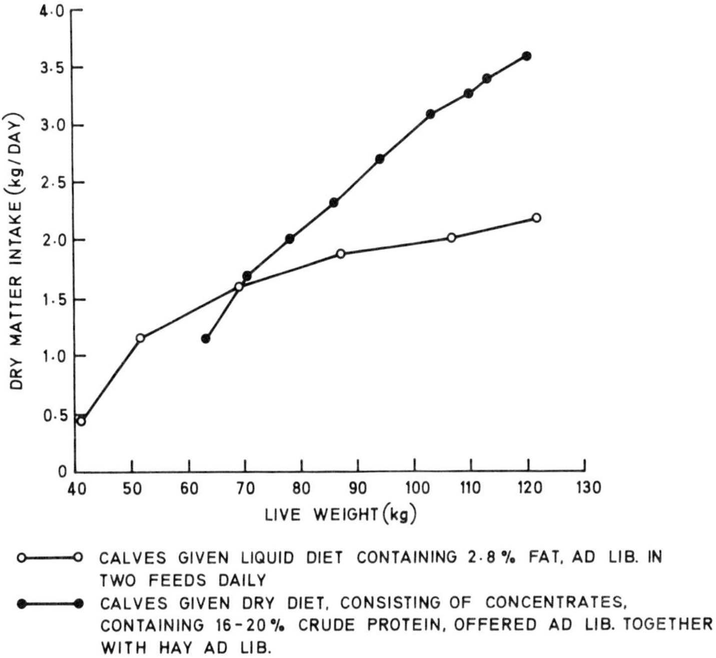

Fig. 2.1. Maximum appetite for dry matter in relation to live weight

efficiency of food conversion, commercial producers of veal calves (see Volume 1, p. 61) sometimes use reconstitution rates of 20–25% in the final stages of fattening, because they claim that this gives a better finish to the carcase. With liquid diets, no difference has been found between the Jersey, Ayrshire and Friesian breeds in the intake of dry matter/unit of live weight over the weight range 60–110 kg (132–242 lb).

On solid food, the dry matter intake of calves tends to rise with an increase in the digestible energy content of the diet; thus a calf of a given weight can eat less of a ration containing a high proportion of fibre, than of one that is more concentrated. This is due in part to the lower digestibility of a fibrous ration (see Table 1.1.) and also to the slower rate of passage of the food resulting in distension of the digestive tract, particularly the reticulo-rumen; concentrates tend to pass rapidly from the rumen and through the digestive tract.[2–4]

Appetite is also affected by the physical environment of the calf. Excessively high temperatures will reduce appetite and low temperatures will tend to increase it. The effect on appetite of temperature differing within a more normal range has not been defined, although it has been claimed that a low relative humidity may reduce appetite in intensively-fed veal calves. Changes in hormone balance will also produce changes in appetite; on the day prior to and on the day of oestrus, food intake is reduced, whereas intake increases on the 2nd–3rd day after oestrus.[5]

WATER

Water consumption is affected not only by the dry matter intake and environmental temperature, but also by the proportion of salt[6] and protein[7–9] in the diet. At a given live weight, low levels of dry matter intake are generally associated with high levels of water intake/unit of dry matter.[10] Moreover, a high water intake induced by a high protein diet may not be reduced when calves are changed over to a lower protein diet.[11] If water is offered to calves given only liquid diets, the amount of water consumed will be dependent on the dry matter content of the liquid diet. With a liquid diet of 5% dry matter the amount of additional water consumed will be negligible; with higher contents of dry matter in the diet, the amount of water consumed will increase. Thus, the total intake of water (from the diet plus additional water) with a diet of 5% dry matter was found to be 18·7 kg water/kg dry matter compared with 4·6 kg dry matter with a diet containing 25% dry matter.[1] The most efficient utilisation of dry matter was obtained with 6·4 kg water/kg dry matter. Under practical conditions, calves given liquid diets only, for veal production are not offered additional water, but it may be advisable to offer water if an outbreak of diarrhoea occurs.

Water intake of ruminant calves in an environmental temperature of up to 10°C (50°F) remains constant at about 3·1–3·5 kg/kg dry matter, but at a temperature of 21·1°C (70°F) the value is about 4·4 kg/kg dry matter and at a temperature of 29·4°C (85°F) is about 5·9 kg/kg dry matter.[12,13]

Where weaning at a young age (3–5 weeks) is practised, clean fresh water should be available when concentrates are first offered, i.e. at 1 week of age. Where weaning is delayed and larger quantities of milk are given, water need not be given until 3 weeks of age. When calves are reared on liberal quantities of skim milk, together with concentrates and hay, much higher intakes of dry food, particularly of hay, will be achieved if water is given *ad lib.*

ENERGY

The energy requirement of the calf, as for other animals, may be subdivided into that required for maintenance and that required for growth.

Of the gross energy (heat of combustion) of a particular diet, energy is lost in the faeces and also in the urine, and, in the case of the ruminant calf, an additional amount is lost as energy in the gases produced as a result of fermentation in the rumen. The gross energy of the food less the energy lost from these three sources is known as the metabolisable energy.

In the ruminant calf, heat is also produced in the fermentation process, and if this value is deducted from the metabolisable energy, the resultant value is called the corrected metabolisable energy.

From the metabolisable energy, or corrected metabolisable energy in the case of the ruminant calf, there are three further losses of energy as heat. First, heat is produced as a result of oxidation of body tissue to provide energy for respiration, circulation, muscle tonus or other vital processes. This loss of heat is known as the basal, or in the ruminant calf, the fasting metabolism. Secondly, heat is produced as a result of feeding (the heat increment of feeding), due to the metabolic processes that are occurring within the tissues. Finally, heat is produced as a result of normal voluntary activity such as drinking, walking, standing up and lying down, and play.

The metabolisable energy, or corrected metabolisable energy, less these three heat losses gives the net energy stored, mainly as fat and protein, which results in growth of the calf.

It will be appreciated that in the pre-ruminant calf the utilisation of food will be much more efficient than in the ruminant calf. First, the losses of energy in the faeces of the pre-ruminant calf are much lower; the digestibility of the dry matter of milk is about 95%, i.e. only 5% of the dry matter consumed appears in the faeces, compared with the highest value for digestibility of concentrates of about 82%. Secondly, no energy losses occur as a result of fermentation in the pre-ruminant calf.

In making full use, at an early age, of the rumen, with its capacity for digesting cheap foods, the lower digestibility of such foods compared with that of milk and the great loss of energy that occurs in the fermentation process must be put on the debit side. The calf reared on dry food from an early age thus has to ingest more energy than a similar animal reared on milk if the same weight gains are to be achieved.

REQUIREMENT FOR MAINTENANCE

The maintenance requirement of the calf consists of the energy utilised for its basal or fasting metabolism, including a small loss of energy in the urine, together with the heat produced by voluntary activity.

In the pre-ruminant calf, the basal metabolic rate is measured with the calf at rest and when the post-absorptive state after feeding has been reached, usually about 15 hr after the last meal. The calf must also be kept in a thermoneutral environment, i.e. above the critical temperature, below which the calf has to oxidise tissue, particularly fat, to keep itself warm. In the ruminant calf, it is difficult to reach the post-absorptive state because of the high gut fill of the rumen and the length of time that food is retained in the alimentary tract; thus, in such calves basal metabolic rate is usually measured on the 4th–5th day after the last feed, and is known as the fasting metabolism.

The basal or fasting metabolism of young calves is considerably greater per unit of body weight or per unit of surface area than that of the adult ruminant. It is probable that the basal metabolic rate of the calf increases during the first 2 days of life, possibly as the result of its effort to adapt itself to its new environment, since the body temperature of the calf does not reach its highest value, which is greater than that for an adult cow, until the second day of life. Thereafter, the basal metabolism per unit of surface area or per kilogramme weight$^{0\cdot73}$ ($W^{0\cdot73}$) declines with age, rapidly during the first 2 weeks of life and then more slowly.[14–16] In addition to this slow decline with age, there appears to be a more marked decline with increasing weight of the animal.

Values for basal metabolism varying from 120 kcal/$W^{0\cdot73}$ to 150 kcal/$W^{0\cdot73}$ have been reported for calves at 1 month of age.[17–21] Standards for fasting metabolism that have recently been accepted[22] until further information is available are as follows:

Age (months)	Fasting metabolism (kcal/kg$^{0\cdot73}$)
1	140
3	135
6	125
12	110

However, it seems probable that these values are considerably lower when calves have been reared on a high plane of nutrition so that they have reached a high percentage of their mature weight at a young age. Indeed, recent work in the Netherlands[23] has shown that the basal metabolic rate for calves given milk substitute *ad lib.* averages 109·4 kcal/$W^{0\cdot75}$ during the age period 15–118 days and for a weight range of 43–153 kg (95–337 lb).

Differences between breeds in fasting metabolism have not been clearly established. In the U.S.A., it has been suggested that at a given age over 6 months, Holstein cattle have a higher fasting metabolism/unit of surface area than Jersey cattle, whereas at a given weight up to 100 kg (220 lb) live weight, fasting metabolism appeared to be higher for the

Jersey breed.[24] In the U.S.S.R., a comparison of the basal metabolism of beef and dairy calves showed that the values for the former were 10% lower at 2 months of age and 19% lower at 12 months.[25]

Heat production has been shown to be closely related to pulse rate, which falls from a value of about 130 beats/min at birth to about 80–90 beats/min at 3 weeks of age.[14,20] Moreover, both the heat production and the pulse rate are increased at higher planes of nutrition.

The heat lost as a result of energy expenditure from body movement is considered to be not more than 10% of the fasting metabolism, when calves are grouped in a pen and given a high plane of nutrition.[21] On lower planes of nutrition, it is likely to be less, as it has been observed that play activity of calves increases with increasing plane of nutrition.[26]

In the pre-ruminant calf, the efficiency of utilisation, or net availability, of the metabolisable energy for maintenance purposes is between 79·5% and 84·5%.[15,27] For the ruminant calf, efficiency is considerably lower and depends on the concentration of energy in the ration. With high-roughage diets containing 2·1 Mcal metabolisable energy/kg dry matter, efficiency is about 68%, whereas with high concentrate diets containing 3·4 Mcal metabolisable energy/kg dry matter, the efficiency is increased to about 78%.[22]

REQUIREMENT FOR GROWTH

The energy storage in the form of fat and protein for a gain in weight of 1 kg/day (2·2 lb/day) is about 2 500 kcal at 100 kg (220 lb) live weight and below, increasing to about 4 000 kcal storage at 400 kg (880 lb) live weight (for details of carcase composition, see Volume 1, p. 62).

The net availability of metabolisable energy (i.e. the percentage of the metabolisable energy that is stored) for the growth of the pre-ruminant calf is very similar to that for maintenance, i.e. between 79·5 and 84·5%, but at very high levels of feed intake, the efficiency probably declines. In the Netherlands, with *ad lib.* feeding of milk substitute, the net availability was only 66% during the period 15–118 days for calves of a weight range from 43 to 153 kg (95 to 337 lb).[23] Total daily heat production for calves given high levels of milk substitute could be estimated by the following two formulae:

$$\text{Heat production (kcal)} = 72{\cdot}0\, W^{0{\cdot}75} + 1\,538\, F$$
$$\text{or} \quad = 18\, W + 1\,538\, F + 454$$

Where W is the live weight (kg) and F is the weight of milk substitute powder (94·8% dry matter) (kg).
With *ad lib.* liquid diets, only a slight fall in digestibility occurs.[28]

With the ruminant calf, the efficiency of utilisation of metabolisable energy (uncorrected) for growth is much lower than that for maintenance. Moreover, a marked fall off in efficiency occurs with high roughage diets, which is even greater at high levels of feeding. The efficiency of utilisation for a high roughage diet containing 2·1 Mcal metabolisable energy/kg dry matter is only 42% compared with 58% for a high concentrate diet containing 3·0 Mcal/kg dry matter.

ENERGY STANDARDS FOR CALVES

In view of the differences in the energy losses between the pre-ruminant and ruminant calf, approximate energy requirements for the two types of calves are given separately in Tables 2.1 and 2.2, respectively.

The Pre-ruminant Calf

In Table 2.1, a maintenance requirement of 52 kcal digestible energy/kg live weight is assumed.[29] This is an approximation because, although the requirement of the young calf in terms of net energy has been found to be about 45 kcal/kg empty body weight,[29–31] it is clear that in the first 10 days of life the value will be considerably higher and may well be lower above 120 kg (265 lb) live weight.

The requirement for gain in weight has been assumed to be 3 000 kcal digested energy/kg weight gain, a value in keeping with the results of research.[29,32,33] As there are only small losses of energy in the urine, metabolisable energy will be about 98% of the digested energy.[33]

The standards are given for whole milk containing 12·6% dry matter and 5·7 Mcal digestible energy/kg dry matter, and for a 20% fat milk substitute containing 5·0 Mcal digestible energy/kg dry matter. The milk substitute containing 96% dry matter is assumed to have been reconstituted at a rate of 1 kg of powder to 6 kg of water to give a dry matter content (as fed) of 13·8%.

Readers wishing to calculate volumes of liquid to feed when using milk substitutes of a different energy content or at different reconstitution rates will need to recalculate the dry matter intake and total weight of liquid required.

The Ruminant Calf

The standards for the ruminant calf are given in terms of crude metabolisable energy (kcal), net energy (kcal) and starch equivalent (lb) (Table 2.2). This has been done because crude metabolisable energy is likely to be the system that will be adopted in this country in the future. For further details of the system see reference 22. Crude metabolisable energy has been converted to starch equivalent (s.e.). The values for cattle of 200, 300 and 400 kg (440, 660 and 880 lb) live weight have been obtained from reference 22. The values for calves under 200 kg have been obtained from results at Shinfield[34] from which it would appear that the accepted fasting metabolism values, given on p. 22, are too high. Moreover, it must be remembered that the values for live weight at the same empty body weight are much higher for the ruminant calf than for the pre-ruminant calf, because of differences in gut fill.

PROTEIN

The protein requirements of the calf have been estimated from feeding trials, from nitrogen balance studies to ascertain the minimum intake that gives maximum retention or, alternatively, by the factorial method.[22]

ENDOGENOUS URINARY N

If a calf is given a nitrogen-free diet, there is still a loss of nitrogen in the urine and faeces. This loss in the urine as a result of metabolic processes in the tissues is called the endogenous urinary N. The endogenous urinary N per unit of body weight, like the fasting metabolism, declines with age. With pre-ruminant calves, the value is between 63 and 82 mg N/kg body weight,[19,33,35,36] and with ruminant calves values decreasing from 58 mg N/kg body weight during the first 12 weeks of life to 25 mg N/kg at a live weight of 283 kg (624 lb) have been reported.[37-41]

METABOLIC FAECAL N

Similarly, there is an inevitable loss of nitrogen in the faeces arising from the digestive juices, bacterial residues and epithelial cells that become eroded during the passage of food through the digestive tract. This loss of nitrogen in the faeces is called metabolic faecal nitrogen. The amount of the loss depends on the dry matter intake and the amount of fibre in the ration. In pre-ruminant calves, the amount of metabolic

Table 2.1 DAILY REQUIREMENT OF THE PRE-RUMINANT CALF IN TERMS OF MATTER) AND A 20% FAT MILK SUBSTITUTE (RECONSTITUTED TO GIVE 13·8%

52 kcal DIGESTIBLE ENERGY/KG LIVE WEIGHT FOR MAINTENANCE, 3 000 kcal

Live weight (kg)	Live weight (lb)	Appetite for dry matter from liquid diets (kg)	Appetite for dry matter from liquid diets (lb)	Maintenance: Digestible energy (kcal)	Whole milk: Dry matter (kg)	Whole milk: Liquid (kg)	Whole milk: Liquid (lb)	Milk substitute: Dry matter (kg)	Milk substitute: Liquid (kg)	Milk substitute: Liquid (lb)
20	44	0·5	1·1	1 040	0·2	1·4	3	0·2	1·5	3
30	66	0·8	1·8	1 560	0·25	2·2	5	0·3	2·3	5
40	88	1·0	2·2	2 080	0·35	2·9	6	0·4	3·0	7
50	110	1·3	2·9	2 600	0·45	3·6	8	0·5	3·8	8
60	132	1·4	3·1	3 120	0·55	4·3	10	0·6	4·5	10
70	154	1·6	3·5	3 640	0·65	5·1	11	0·75	5·3	12
80	176	1·8	4·0	4 160	0·75	5·7	13	0·85	6·0	13
90	198	1·9	4·2	4 680	0·8	6·5	14	0·95	6·8	15
100	220	1·95	4·3	5 200	0·9	7·2	16	1·05	7·5	17
110	243	2·05	4·5	5 720	1·0	8·0	18	1·15	8·3	18
120	265	2·15	4·7	6 240	1·1	8·7	19	1·25	9·0	20
130	287	2·20	4·9	6 760	1·2	9·4	21	1·35	9·8	22
140	309	2·25	5·0	7 280	1·3	10·1	22	1·45	10·6	23

Table 2.1 *continued*

Live weight (kg)	Live weight (lb)	Appetite for dry matter from liquid diets (kg)	Appetite for dry matter from liquid diets (lb)	Maintenance + 1·0 kg gain/day (2·2 lb/day): Digestible energy (kcal)	Whole milk: Dry matter (kg)	Whole milk: Liquid (kg)	Whole milk: Liquid (lb)	Milk substitute: Dry matter (kg)	Milk substitute: Liquid (kg)	Milk substitute: Liquid (lb)
20	44	0·5	1·1	–	–	–	–	–	–	–
30	66	0·8	1·8	4 560	0·8	6·3	14	0·9	6·6	15
40	88	1·0	2·2	5 080	0·9	7·1	16	1·0	7·4	16
50	110	1·3	2·9	5 600	1·0	7·8	17	1·1	8·1	18
60	132	1·4	3·1	6 120	1·05	8·5	19	1·2	8·9	20
70	154	1·6	3·5	6 640	1·15	9·2	20	1·3	9·6	21
80	176	1·8	4·0	7 160	1·25	10·0	22	1·45	10·4	23
90	198	1·9	4·2	7 680	1·35	10·7	24	1·55	11·1	25
100	220	1·95	4·3	8 200	1·45	11·4	25	1·65	11·9	26
110	243	2·05	4·5	8 720	1·55	12·1	27	1·75	12·6	28
120	265	2·15	4·7	9 240	1·6	12·9	28	1·85	13·4	30
130	287	2·20	4·9	9 760	1·7	13·6	30	1·95	14·1	31
140	309	2·25	5·0	10 280	1·8	14·3	32	2·05	14·9	33

WHOLE MILK (12·6% DRY MATTER, 5·7 kcal DIGESTIBLE ENERGY/g DRY
DRY MATTER, 5·0 kcal DIGESTIBLE ENERGY/g DRY MATTER).

DIGESTIBLE ENERGY FOR 1 kg GAIN IN WEIGHT. (1 kg = 2·2 lb)

Maintenance + 0·25 kg gain/day (0·55 lb/day)							*Maintenance + 0·50 kg gain/day (1·1 lb/day)*						
	Whole milk			*Milk substitute*				*Whole milk*			*Milk substitute*		
Digestible energy (kcal)	*Dry matter (kg)*	*Liquid (kg)*	*(lb)*	*Dry matter (kg)*	*Liquid (kg)*	*(lb)*	*Digestible energy (kcal)*	*Dry matter (kg)*	*Liquid (kg)*	*(lb)*	*Dry matter (kg)*	*Liquid (kg)*	*(lb)*
1 790	0·3	2·5	6	0·35	2·6	6	2 540	0·45	3·5	8	0·5	3·7	8
2 310	0·4	3·2	7	0·45	3·3	7	3 060	0·55	4·3	10	0·6	4·4	10
2 830	0·5	3·9	9	0·55	4·1	9	3 580	0·65	5·0	11	0·7	5·2	12
3 350	0·6	4·7	10	0·65	4·9	11	4 100	0·7	5·7	13	0·8	5·9	13
3 870	0·7	5·4	12	0·75	5·6	12	4 620	0·8	6·4	14	0·9	6·7	15
4 390	0·75	6·1	13	0·9	6·4	14	5 140	0·9	7·2	16	1·05	7·4	16
4 910	0·85	6·8	15	0·95	7·0	15	5 660	1·0	7·9	17	1·15	8·2	18
5 430	0·95	7·6	17	1·1	7·9	17	6 180	1·1	8·6	19	1·25	9·0	20
5 950	1·05	8·3	18	1·2	8·6	19	6 700	1·2	9·3	21	1·35	9·7	21
6 470	1·15	9·0	20	1·3	9·4	21	7 220	1·25	10·0	22	1·45	10·5	23
6 990	1·25	9·7	21	1·4	10·1	22	7 740	1·35	10·8	24	1·55	11·2	25
7 510	1·30	10·5	23	1·5	10·9	24	8 260	1·45	11·5	25	1·65	12·0	27
8 030	1·40	11·2	25	1·6	11·6	26	8 780	1·55	12·2	27	1·75	12·7	28

Maintenance + 1·5 kg gain/day (3·3 lb/day)						
	Whole milk			*Milk substitute*		
Digestible energy (kcal)	*Dry matter (kg)*	*Liquid (kg)*	*(lb)*	*Dry matter (kg)*	*Liquid (kg)*	*(lb)*
–	–	–	–	–	–	–
–	–	–	–	–	–	–
–	–	–	–	–	–	–
7 100	1·25	9·9	22	1·4	10·3	23
7 620	1·35	10·6	23	1·5	11·0	24
8 140	1·45	11·3	25	1·6	11·8	26
8 660	1·5	12·1	27	1·75	12·5	28
9 180	1·6	12·8	28	1·85	13·3	29
9 700	1·7	13·5	30	1·95	14·1	31
10 220	1·8	14·2	31	2·05	14·8	33
10 740	1·9	14·9	33	2·15	15·6	34
11 260	2·0	15·7	35	2·25	16·3	36
11 780	2·05	16·4	36	2·35	17·1	38

Table 2.2 DAILY REQUIREMENT OF THE RUMINANT CALF IN TERMS OF METABOLISABLE ENERGY (m.e.), NET ENERGY AND STARCH EQUIVALENT (s.e.)

				Maintenance (*2·6 Mcal m.e./kg dry matter, e.g. 50% hay, 50% concentrates*)			*Maintenance + 0·50 kg gain/day (1·1 lb/day) (3·0 Mcal m.e./kg dry matter, e.g. 15% hay, 85% concentrates)*			*Maintenance + 1·0 kg gain/day (2·2 lb/day) (3·1 Mcal m.e./kg dry matter, e.g. 9% hay, 91% concentrates)*			*Maintenance + 1·5 kg gain/day (3·3 lb/day) (3·2 Mcal m.e./kg dry matter, e.g. 100% concentrates)*		
Live weight (kg)	*Live weight (lb)*	*Appetite for dry matter (kg)*	*Appetite for dry matter (lb)*	*Metabolisable energy (kcal)*	*Net energy (kcal)*	*s.e. (lb)*	*Metabolisable energy (kcal)*	*Net energy (kcal)*	*s.e. (lb)*	*Metabolisable energy (kcal)*	*Net energy (kcal)*	*s.e. (lb)*	*Metabolisable energy (kcal)*	*Net energy (kcal)*	*s.e. (lb)*
50	110	1·0	2	2 900	2 100	1·4	4 500	3 050	2·4	6 600	4 200	3·7	9 300	5 850	5·4
75	165	1·8	4	3 600	2 600	1·7	5 200	3 550	2·8	7 400	4 800	4·2	10 500	6 550	6·1
100	220	3·0	7	4 400	3 150	2·1	6 000	4 200	3·3	8 400	5 550	4·7	11 400	7 400	6·6
125	276	3·8	8	5 100	3 750	2·4	6 800	4 750	3·7	9 500	6 250	5·3	12 600	8 200	7·3
150	330	4·5	10	5 900	4 250	2·8	7 600	5 300	4·1	10 200	6 900	5·7	13 600	8 950	7·9
175	386	5·3	12	6 600	4 750	3·2	8 300	5 850	4·5	11 200	7 550	6·3	14 700	9 700	8·5
200	441	6·0	13	8 000	5 800	3·6	9 900	6 950	5·4	12 750	8 700	7·2	16 500	11 000	9·6
300	661	9·0	20	9 800	7 050	4·7	12 100	8 450	6·6	15 350	10 500	8·6	20 050	13 350	11·6
400	882	12·0	26	11 500	8 300	5·5	14 100	9 150	7·7	17 950	12 300	10·1	23 850	15 800	13·8

faecal N is, as might be expected, considerably lower than for the ruminant calf. The value for the pre-ruminant calf given whole milk or a good quality milk substitute is about 1·9 g N/kg dry matter intake,[33] although higher values have been reported,[19,35,42] whereas for the ruminant calf given a dry diet containing 30% crude fibre the value is about 4·2 g N/kg dry matter and for one containing only 5% crude fibre the value is about 3·3 g N/kg dry matter.[9,37]

N CONTENT OF WEIGHT GAIN

The amount of nitrogen stored for each kilogramme gain in weight is generally between 26 and 34 g,[9,29,32,33,43] but possibly due to differences in carcase composition, it appears to be higher for calves gaining weight very slowly, and also for calves gaining weight very rapidly if the diet contains a high concentration of protein.

DIGESTIBILITY OF PROTEIN

The apparent digestibility of the protein of whole milk or a good quality milk substitute is about 97 and 94%, respectively. Most if not all of the nitrogen loss in the faeces on whole milk diets is of metabolic origin, so the true digestibility of the protein approaches 100%. For dry diets, the apparent digestibility of protein may be as high as 80% for concentrates, but only 40% for hay.[44] Moreover, the higher the protein content of the diet, the higher is the apparent digestibility of the protein.[37,43]

BIOLOGICAL VALUE OF PROTEIN

Of the protein that is truly digested, only a proportion is retained in the body as protein. This proportion is known as the biological value of the protein. In the pre-ruminant calf, which resembles the single stomached (monogastric) animal, the biological value reaches its maximum when the supply of amino-acids from the protein are in exactly the correct proportions for the requirement of the calf. However, the maximum value for biological value will only be reached when the diet being given to the calf is limiting in protein (i.e. there is an excess of energy and all other nutrients except protein).

Under these conditions the biological value of milk protein is 80%, or even as high as 90%.[32,45] Similarly for dry diets, a mean biological value of 82% was found with protein-limiting diets varying from all-hay to *ad lib.* concentrates containing 12% crude protein (air-dry basis), over a range of weight gains from maintenance to 1·5 kg/day (3·3 lb/day).[37] Due to bacterial synthesis of protein in the rumen, the quality of proteins in dry foods has until recently been considered of little importance for the young ruminant calf (see following sub-section).

PROTEIN REQUIREMENTS

The requirement of available protein (R_{AP}), estimated by the factorial method is obtained from the following equation:[46]

$$R_{AP} = \frac{1}{(BV)} 6{\cdot}25\,(E + G)$$

where BV is the biological value (as a coefficient), E is the endogenous urinary N (g/day), and G is the N retention (g/day) for a particular weight gain. In terms of apparently digested crude protein (R_{ADP}). the requirement is as follows:

$$R_{ADP} = 6{\cdot}25 \left[\frac{1}{(BV)} (E + G + M.D) - M.D \right]$$

where M is the metabolic faecal N excretion (g/kg dry matter ingested), and D is the dry matter intake, in kilogrammes, required to supply the energy needs of the calf.

To convert from available protein requirement, which is independent of the dry matter intake of the calf, to digestible crude protein, it is only necessary to add on the factor $6{\cdot}25\,[(M.D/BV) - M.D]$ to the available protein requirement.

Table 2.3 gives the requirement of available protein of calves. As can be seen, the requirement of digestible protein is only slightly higher than that of available protein for the pre-ruminant calf, but more markedly higher for the ruminant calf. This is due partly to the higher dry matter intake necessary with dry food to achieve the same weight gain and partly to the higher metabolic faecal N of the ruminant calf. Table 2.4 shows the minimum content of crude protein required in milk substitute or concentrate diets for calves of different live weights gaining weight at different rates. From this table, the following points are evident. First, a milk substitute must contain a higher content of protein than a dry concentrate; secondly, the greater the weight gain required, the higher must be the content of protein; and thirdly, the heavier the calf the lower the protein content required for a given weight gain. It must be appreciated that the values of protein content given for concentrate-fed calves refer to calves given a diet solely of concentrates. The limiting factor for growth in the calf weaned on to dry food at an early age is the amount of energy consumed in the diet, which in turn is associated largely with the palatability of the diet. If it is intended to restrict the concentrate intake of calves, it will be necessary to feed a concentrate of slightly higher protein content, depending on the protein content of the hay. However, restriction of concentrates will result in a lower growth rate, which in turn will reduce the protein requirement of the calf.

The values for the protein content of diets are in agreement with experimental observations. In whole milk about 26% of the calories are derived from protein and, when sufficient energy is given to cover the maintenance requirement, as little as 7% of the calories are necessary as protein, whereas for a gain of 1 kg/day (2·2 lb/day), the value is about 22%.[29] These values are for milk protein that is completely digested and retained in the body, so that the actual percentages required are somewhat higher. Similarly, it is known that a milk substitute containing 19·3% crude protein will not sustain maximum growth in calves of 80 kg (176 lb) live weight when the diet is given *ad lib.*[47]

For the ruminant calf weaned at 5 weeks of age, it is known that a concentrate mixture containing 16% crude protein (air dry basis), even when restricted to a level of 2 kg/day (4·4 lb/day) will supply the needs of the calf.[11,38,48] Moreover, if concentrates are offered *ad lib.*, a mixture containing 12% crude protein (air dry basis) will supply the needs of Friesian calves after 8 weeks of age, i.e. at about 70 kg (154 lb) live weight.[9] As this is the minimum content of protein required, it is suggested that on an *ad lib.* system of feeding concentrates, the mixture should contain 14% crude protein (air dry basis) after 8 weeks of age (see also Volume 1, p. 82).

QUALITY OF PROTEIN

The importance of the quality of the protein included in concentrate mixtures for calves is still a matter of debate. There is some evidence that, in spite of bacterial synthesis of protein and B vitamins in the rumen, the biological value of vegetable protein is lower than for animal protein and that both solubility and amino-acid content, particularly of lysine and methionine, are of importance in the ruminant calf; a biological value of 70% was reported for groundnut meal compared with 83% for white fish meal, given to calves weaned at 3 weeks of age.[49]

A considerable wastage of nitrogen occurs in the ruminant, owing to the absorption of ammonia produced in the rumen, which is subsequently excreted in the urine as urea. If the solubility of the protein is reduced by heat treatment, this loss of nitrogen can be reduced especially on high-concentrate diets.[50] Moreover, on such diets, it is suggested that the protein may escape degradation in the rumen and pass into the

Table 2.3 DAILY REQUIREMENT OF THE PRE-RUMINANT CALF FOR AVAILABLE AND APPARENTLY DIGESTIBLE CRUDE PROTEIN

Live weight		*Maintenance*					*Maintenance + 0·50 kg gain/day (1·1 lb/day)*				
		Dry matter intake (kg)		*Available protein*	*Digestible protein (g)*		*Dry matter intake (kg)*		*Available protein*	*Digestible protein (g)*	
(kg)	*(lb)*	*Pre-ruminant*	*Ruminant*	*(g)*	*Pre-ruminant*	*Ruminant*	*Pre-ruminant*	*Ruminant*	*(g)*	*Pre-ruminant*	*Ruminant*
20	44	0·2	–	15	15	–	0·5	–	130	130	–
40	88	0·4	0·8	20	25	25	0·7	1·4	140	140	145
60	132	0·6	0·9	30	30	35	0·9	1·6	145	150	155
80	176	0·8	1·1	35	40	40	1·1	1·8	155	155	160
100	220	1·0	1·3	40	45	50	1·3	2·0	160	165	170
120	265	1·3	1·5	45	50	55	1·6	2·1	165	170	175
140	309	1·5	1·6	55	55	60	1·8	2·3	170	175	180

For the calculation of the protein requirements the following factors have been used: Endogenous urinary nitrogen, 184·4 mg/$W^{0·73}$. N-retention, 30 g N kg gain in weight. Biological value, 80%. Metabolic faecal N; pre-ruminant calf, 1·9 g/kg dry matter intake, ruminant calf, 3·3 g/kg dry matter intake (concentrates)

200	441	–	–	60	–	–	–	–	170	–	–
300	661	–	–	80	–	–	–	–	185	–	–
400	882	–	–	100	–	–	–	–	210	–	–

The values of available protein for 200–400 kg live weight are obtained from reference 22. To convert to digestible crude protein, add 13·4 g protein/kg dry matter intake.

Table 2.3 *continued*

Live weight		*Maintenance + 1·0 kg gain/day (2·2 lb/day)*					*Maintenance + 1·5 kg gain/day (3·3 lb/day)*				
		Dry matter intake (kg)		*Available protein*	*Digestible protein (g)*		*Dry matter intake (kg)*		*Available protein*	*Digestible protein (g)*	
(kg)	*(lb)*	*Pre-ruminant*	*Ruminant*	(g)	*Pre-ruminant*	*Ruminant*	*Pre-ruminant*	*Ruminant*	*(g)*	*Pre-ruminant*	*Ruminant*
20	44	0·8	–	245	250	–	–	–	–	–	–
40	88	1·0	–	255	260	–	–	–	–	–	–
60	132	1·2	2·4	265	270	275	1·5	–	380	385	–
80	176	1·4	2·6	270	275	285	1·7	3·2	385	390	405
100	220	1·6	2·7	275	280	290	1·9	3·4	395	400	410
120	265	1·9	2·9	280	285	295	2·2	3·5	400	405	415
140	309	2·1	3·1	290	295	305	2·4	3·7	405	410	425

For the calculation of the protein requirements the following factors have been used: Endogenous urinary nitrogen, 184·4 mg/$W^{0.73}$. N-retention, 30 g N kg gain in weight. Biological value, 80%. Metabolic faecal N; pre-ruminant calf, 1·9 g/kg dry matter intake, ruminant calf, 3·3 g/kg dry matter intake (concentrates)

200	441	–	–	270	–	–	–	–	–	–	–
300	661	–	–	290	–	–	–	–	–	–	–
400	882	–	–	310	–	–	–	–	–	–	–

The values of available protein for 200–400 kg live weight are obtained from reference 22. To convert to digestible crude protein, add 13·4 g protein/kg dry matter intake.

Table 2.4 MINIMUM CRUDE PROTEIN CONTENT (%) REQUIRED IN MILK SUB-PROTEIN: 94% FOR MILK SUBSTITUTE, 71% FOR CONCENTRATES)

Live weight		*Maintenance*				*Maintenance + 0·50 kg/day (1·1 lb day)*			
		Milk substitute		*Concentrates*		*Milk substitute*		*Concentrates*	
(kg)	*(lb)*	*air-dry basis (96% DM)*	*Dry matter basis*	*air-dry basis (85% DM)*	*Dry matter basis*	*air-dry basis (96% DM)*	*Dry matter basis*	*air-dry basis (85% DM)*	*Dry matter basis*
20	44	6·5	6·5	–	–	26·5	27·5	–	–
40	88	5·5	6·0	4·0	4·5	20·0	21·0	12·0	14·5
60	132	5·0	5·0	4·0	5·0	16·5	17·0	11·5	13·5
80	176	4·5	5·0	4·5	5·0	14·0	14·5	11·0	13·0
100	220	4·5	4·5	4·5	5·5	12·5	13·0	10·5	12·0
120	265	4·0	4·5	4·5	5·5	11·0	11·5	10·0	11·5
140	309	4·0	4·0	4·5	5·5	10·0	10·5	9·5	11·0

abomasum; from then onwards quality of protein becomes as important as in the monogastric animal. No difference was found between N-retention in calves given two groundnut meals of markedly different solubilities at pH 7, but a difference was found for two herring meals, the 'insoluble' herring meal giving a higher N-retention than the 'soluble' herring meal. These divergent results were attributed to the fact that at pH 4, a value somewhat lower than a pH of 5–5·5 which may occur in the rumen of calves given high-concentrate diets, there was no difference in the solubility of the two groundnut meals, but the solubility of the herring meal was unaffected by pH. Moreover, the higher N-retention obtained from 'insoluble' herring meal, compared with that from groundnut meals, was attributed to the higher lysine and methionine content of the herring meal.[51]

Nevertheless, satisfactory growth rates at restricted levels of concentrate intake have been reported when simple concentrate mixtures containing only one major source of protein, often of vegetable origin, have been compared with more complex mixtures containing a variety of protein-rich ingredients, irrespective of the age at weaning.[52–59]

The inclusion of milk protein in concentrate mixtures does not appear to be necessary, but slightly improved growth rates were obtained with Ayrshire calves weaned at 3 weeks of age, when linseed meal, included at 10% of the diet, was replaced by an equal quantity of dried skim milk.[60]

Because of the danger of poisoning from gossypol, present in cottonseed cake, it is unwise to feed this cake to calves under 3–4 months of age. However, cottonseed meal containing only about 0·029% free

TITUTE AND CONCENTRATE DIETS FOR CALVES (DIGESTIBILITY OF CRUDE

Maintenance +1·0 kg gain/day (2·2 lb/day)				*Maintenance + 1·5 kg gain/day (3·3 lb/day)*			
Milk substitute		*Concentrates*		*Milk substitute*		*Concentrates*	
air-dry basis 96% DM)	*Dry matter basis*	*air-dry basis (85% DM)*	*Dry matter basis*	*air-dry basis (96% DM)*	*Dry matter basis*	*air-dry basis (85% DM)*	*Dry matter basis*
–	–	–	–	–	–	–	–
26·0	27·0	–	–	–	–	–	–
22·5	23·5	14·0	16·5	26·0	27·0	–	–
19·5	20·5	13·5	15·5	23·0	24·0	15·0	18·0
17·0	17·5	13·0	15·0	21·0	22·0	14·5	17·0
16·0	16·5	12·5	14·5	19·0	20·0	14·0	16·5
14·5	15·0	12·0	14·0	18·0	18·5	14·0	16·0

gossypol could form up to 60% of a concentrate mixture without ill effect.[61]

NON-PROTEIN NITROGEN

Experiments on the use of urea in calf diets have been made as long ago as 1943,[62] when it was shown with 2-month old calves that a diet containing 4·4% protein equivalent would not support growth, but the addition of sufficient urea to raise the protein equivalent to 16·2% resulted in a growth rate of 0·45 kg/day (1 lb/day) between 2 and 4 months of age. On addition of urea, the apparent digestibility of the dry matter and carbohydrates of the diet were markedly increased. Similarly, calves weaned at 7 weeks of age and given a urea supplement to a basal mixture containing 6·7% crude protein utilised appreciable quantities of urea to meet their requirements for growth as early as the 6th week of life.[63,64] With calves weaned at 5 weeks of age and given a concentrate mixture containing 12% crude protein (air-dry basis), a slight improvement in performance was achieved[65] by supplementing the diet with urea (33% of the total nitrogen in the diet, i.e. 28 g urea/kg concentrates) during the first 8 weeks of life, but thereafter the urea supplement tended to reduce performance, partly due to its low palatability.

Although there is little doubt that urea can be satisfactorily utilised by the young ruminant calf, its inclusion in concentrate diets for use during the early period of life is debatable, especially if maximum weight gains are required on dry diets.

REFERENCES

1. PETTYJOHN, J. D., EVERETT, J. P. Jr. and MOCHRIE, R. D. *J. Dairy Sci.* **46**, 710 (1963)
2. BLAXTER, K. L., GRAHAM, N. McC. and WAINMAN, F. W. *Br. J. Nutr.* **10**, 69 (1956)
3. PALOHEIMO, L. and MÄKELÄ, A. *Suom. maatal. Seur. Julk.* **94**, 15 (1959)
4. FREER, M. and CAMPLING, R. C. *Br. J. Nutr.* **17**, 79 (1963)
5. RAUN, A. P., McASKILL, J. W., COOLEY, C. O., WAGNER, J. F. and TOMKINSON, L. *J. Anim. Sci.* **26**, 949 (1967)
6. RIGGS, J. K., COLBY, R. W. and SELLS, L. V. *J. Anim. Sci.* **12**, 379 (1953)
7. RITZMAN, E. G. and BENEDICT, F. G. *Tech. Bull. New Hamps. agric. Exp. Stn* no. 26 (1924)
8. SYKES, J. F. *Yb. Agric. U.S. Dep. Agric.* p. 14 (1955)
9. STOBO, I. J. F., ROY, J. H. B. and GASTON, H. J. *Anim. Prod.* **9**, 7 (1967)
10. LEITCH, I. and THOMSON, J. S. *Nutr. Abstr. Rev.* **14**, 197 (1944–45)
11. STOBO, I. J. F., ROY, J. H. B. and GASTON, H. J. *Anim. Prod.* **9**, 23 (1967)
12. WINCHESTER, C. F. and MORRIS, M. J. *J. Anim. Sci.* **15**, 722 (1956)
13. MACDONALD, M. A. and BELL, J. M. *Can. J. Anim. Sci.* **38**, 23 (1958)
14. ROY, J. H. B., HUFFMAN, C. F. and REINEKE, E. P. *Br. J. Nutr.* **11**, 373 (1957)
15. GONZALEZ-JIMENEZ, E. and BLAXTER, K. L. *Br. J. Nutr.* **16**, 199 (1962)
16. SETTLEMIRE, C. J., HIBBS, J. W. and CONRAD, H. R. *J. Dairy Sci.* **47**, 875 (1964)
17. RITZMAN, E. G. and COLOVOS, N. F. *Tech. Bull. New Hamps. agric. Exp. Stn* no. 80 (1943)
18. BLAXTER, K. L. and HOWELLS, A. *Br. J. Nutr.* **5**, 25 (1951)
19. BLAXTER, K. L. and WOOD, W. A. *Br. J. Nutr.* **5**, 11 (1951)
20. BLAXTER, K. L. and WOOD, W. A. *Br. J. Nutr.* **5**, 29 (1951)
21. BLAXTER, K. L. *The Energy Metabolism of Ruminants* (Hutchinson, London 1962)
22. Agricultural Research Council *The Nutrient Requirements of Farm Livestock no. 2 Ruminants: Technical Reports and Summaries* (H.M.S.O. London, 1965)
23. van ES, A. J. H., van WEERDEN, E. J. and FRENS, A. M. *Z. Tierphysiol. Tierernähr. Futtermittelk.* **21**, 25 (1966)
24. BRODY, S. *Bioenergetics and Growth.* (Reinhold, New York, 1945)
25. TKAČEV, I. and TARANENKO, G. A. *Vest. sel.'-khoz. Nauki, Mosk.* **5**, 65 (1963)
26. BROWNLEE, A. *Br. vet. J.* **110**, 48 (1954)
27. BLAXTER, K. L. *Br. J. Nutr.* **6**, 12 (1952)
28. ROY, J. H. B., STOBO, I. J. F. and GASTON, H. J. *Rep. natn. Inst. Res. Dairy.* p. 43 (1965)
29. BLAXTER, K. L. and WOOD, W. A. *Br. J. Nutr.* **5**, 55 (1951)
30. TOMME, M. F. and TARANENKO, G. A. *Dokl. vses. Akad. sel.'-khoz. Nauk* no. 18, p. 36 (1939)
31. ROY, J. H. B., SHILLAM, K. W. G., HAWKINS, G. M. and LANG, J. M. *Br. J. Nutr.* **12**, 123 (1958)
32. BRISSON, G. J., CUNNINGHAM, H. M. and HASKELL, S. R. *Can. J. Anim. Sci.* **37**, 157 (1957)
33. ROY, J. H. B., GASTON, H. J., SHILLAM, K. W. G., THOMPSON, S. Y., STOBO, I. J. F. and GREATOREX, J. C. *Br. J. Nutr.* **18**, 467 (1964)
34. STOBO, I. J. F. *Personal communication* (1968)
35. CUNNINGHAM, H. M. and BRISSON, G. J. *Can. J. Anim. Sci.* **37**, 152 (1957)
36. SHILLAM, K. W. G. and ROY, J. H. B. *Br. J. Nutr.* **17**, 171 (1963)

37. STOBO, I. J. F. 'Studies in the nutrition of young cattle with special reference to rumen development and protein requirement of the early-weaned calf.' *Ph.D. Thesis* (University of Reading, 1964)
38. HARRIS, L. E. and LOOSLI, J. K. *J. Dairy Sci.* **27**, 650 (1944)
39. STEENBOCK, H., NELSON, V. E. and HART, E. B. *Res. Bull. agric. Exp. Stn Univ. Wis.* no. 36 (1915)
40. HART, E. B., HUMPHREY, G. C. and MORRISON, F. B. *J. biol. Chem.* **13**, 133 (1912)
41. SWANSON, E. W. and HERMAN, H. A. *Res. Bull. Mo. agric. Exp. Stn* no. 372 (1943)
42. LOFGREEN, G. P. and KLEIBER, M. *J. Nutr.* **49**, 183 (1953)
43. WHITELAW, F. G., PRESTON, T. R. and NDUMBE, R. D. *Anim. Prod.* **3**, 121 (1961)
44. STOBO, I. J. F., ROY, J. H. B. and GASTON, H. J. *Br. J. Nutr.* **20**, 189 (1966)
45. BLAXTER, K. L. and WOOD, W. A. *Br. J. Nutr.* **6**, 56 (1952)
46. BLAXTER, K. L. and MITCHELL, H. H. *J. Anim. Sci.* **7**, 351 (1948)
47. STOBO, I. J. F., ROY, J. H. B. and GASTON, H. J. *Anim. Prod.* **9**, 267 (1967)
48. BROWN, L. D., LASSITER, C. A., EVERETT, J. P., SEATH, D. M. and RUST, J. W. *J. Dairy Sci.* **41**, 1425 (1958)
49. PRESTON, T. R. *Wld Rev. Nutr. Diet.* **4**, 117 (1963)
50. CHALMERS, M. I., CUTHBERTSON, D. P. and SYNGE, R. L. M. *J. agric. Sci., Camb.* **44**, 254 (1954)
51. WHITELAW, F. G. and PRESTON, T. R. *Anim. Prod.* **5**, 131 (1963)
52. HIBBS, J. W., POUNDEN, W. D. and CONRAD, H. R. *J. Dairy Sci.* **36**, 717 (1953)
53. JACOBSON, N. L., ALLEN, R. S. and BELL, M. R. *J. Anim. Sci.* **10**, 1050 (1951)
54. GARDNER, K. E. *J. Dairy Sci.* **35**, 491 (1952)
55. HOLTER, J. A. *Proc. Cornell Nutr. Conf.* p. 53 (1954)
56. LAMBERT, M. R., JACOBSON, N. L., ALLEN, R. S. and BELL, M. R. *J. Dairy Sci.* **38**, 6 (1955)
57. KESLER, E. M. and WILSON, J. M. *Bull. Pa agric. Exp. Stn* no. 634 (1958)
58. MURLEY, W. R., MOCHRIE, R. D., EDWARDS, J. R. and HOLLON, B. F. *J. Dairy Sci.* **40**, 1390 (1957)
59. HARRISON, H. N., WARNER, R. G., SANDER, E. G., LOOSLI, J. K., SLACK, S. T. and TURK, K. L. *J. Dairy Sci.* **43**, 1084 (1960)
60. PRESTON, T. R. *Proc. Br. Soc. Anim. Prod.* p. 67 (1956)
61. HOLLON, B. F., WAUGH, R. K., WISE, G. H. and SMITH, F. H. *J. Dairy Sci.* **41**, 286 (1958)
62. LOOSLI, J. K. and McCAY, C. M. *J. Nutr.* **25**, 197 (1943)
63. BROWN, L. D., LASSITER, C. A., EVERETT, J. P. and RUST, J. W. *J. Anim. Sci.* **15**, 1125 (1956)
64. BROWN, L. D., JACOBSON, D. R., EVERETT, J. P., SEATH, D. M. and RUST, J. W. *J. Dairy Sci.* **43**, 1313 (1960)
65. STOBO, I. J. F., ROY, J. H. B. and GASTON, H. J. *Anim. Prod.* **9**, 155 (1967)

CHAPTER THREE

The Nutrient Requirements of the Calf—Major Inorganic Elements

When liberal amounts of concentrates and leguminous hays are fed in addition to milk, mineral supplements other than salt are seldom necessary.

CALCIUM AND PHOSPHORUS

When the quantity of milk is restricted and the concentrate mixture contains no fish meal, or meat and bone meal, 2% of chalk and/or sterilised steamed bone flour should be added to the mixture. This is probably an unnecessary precaution when good quality leguminous hay is fed, as it is particularly rich in calcium.

Ample phosphorus will be supplied, even after weaning, if the concentrate mixture contains 10–20% of a protein supplement such as wheat bran, middlings, soya bean meal or linseed cake. If, however, the concentrate mixture consists largely of cereal grains, steamed bone flour should definitely be included.

On hill land there is a possibility that calves receiving no supplementary food may suffer from calcium and phosphorus deficiency resulting in rickets (see p. 77). Furthermore, a deficiency in phosphorus alone may also cause rickets, especially if the calf obtains excessive quantities of calcium at the same time. A deficiency of phosphorus, besides being associated with rickets, results in anorexia (loss of appetite) and even a depraved appetite and the animal thus becomes emaciated. Phosphorus deficiency is more easily produced than calcium deficiency, for whereas the skeleton contains about 99% of the calcium in the body, it contains only about 75–80% of the phosphorus, the remainder being mainly in muscle and nervous tissue. Phosphorus deficiency (aphosphorosis) is well known under range conditions in South Africa and Texas.

CALCIUM:PHOSPHORUS RATIO

It is usually recommended that for maximum utilisation of these minerals, the Ca : P ratio of a ration should be similar to that of bone, i.e. 2·2 : 1 (1·3 CaO:1P_2O_5) or, as more phosphorus than calcium is found in the soft tissues, somewhat less. In fact, calves have grown equally well on rations with Ca:P ratios from 6:1 to 1·2:1 when adequate vitamin D was supplied. As a rule, there should be no need to consider the Ca:P ratio of the diet if the requirements of the animal for these two minerals are fulfilled.

CALCIUM REQUIREMENT

The maintenance requirement of this mineral, as for phosphorus, is often assessed from the endogenous faecal excretion, which represents the inevitable loss by excretion into the gut when diets devoid of calcium and phosphorus are given. Reported values for the daily endogenous faecal calcium losses vary between 9 and 28 mg calcium per kilogramme body weight, with no marked trend associated with the age or weight of the calf[1–5] or with calcium intake.[6] Daily urinary losses appear to be not more than 1 mg/kg body weight.

The true absorption of dietary calcium decreases with age, but is markedly affected by the source of calcium, absorption from milk being much higher than from dry food. The percentage that is truly absorbed from milk in the young calf is at least 92% and may be as high as 98%, but only 83% at 170 kg (374 lb) live weight.[3,7–9] When calcium intake is much in excess of requirements, the percentage absorbed is reduced.[10,11] At a maintenance level of milk intake, the percentage of calcium retained was 73%, but increased to 92% when the amount of milk was sufficient to give a weight gain of 1 kg/day (2·2 lb/day).[1]

The true absorption from dry foods is between 40 and 50%.[3–5] With different sources of calcium in dry food, values for absorption ranged from 41 to 68% for animals of 136–227 kg (300–500 lb) live weight, and from 30 to 55% for those weighing 363–454 kg (800–1 000 lb).[4]

The net requirement of calcium for weight gain has been obtained either from the calcium composition of cattle slaughtered at various ages,[12–15] or from the nitrogen content of the tissues, by assuming a uniform retention of 0·27 g phosphorus to each gramme of nitrogen retained and a calcium retention equal to 1·71 x phosphorus. In the latter case gross requirements were obtained by assuming that 70% of the calcium intake was retained.[16] In general it would appear that, except during the very early period after birth, calcium deposition is about 12 g/kg live weight gain up to 227 kg (500 lb) live weight,[17] although higher values have been used in the recently recommended standards.[18]

The gross requirements for calcium may be obtained by the summation of the endogenous loss in faeces and urine and the net requirement for weight gain, divided by the true absorption as a coefficient. The recently recommended gross requirements of calcium for cattle[18] gaining weight at different rates are given in Table 3.1.

Table 3.1 DAILY CALCIUM, PHOSPHORUS AND MAGNESIUM REQUIREMENT OF CALVES GAINING WEIGHT AT 0·5 kg (1·1 lb) AND 1·0 kg (2·2 lb/day)

Live weight (kg)	Live weight (lb)	*Maintenance + 0·5 kg/day (1·1 lb/day)* Ca (g/day)	P (g/day)	Mg (g/day)	*Maintenance + 1·0 kg/day (2·2 lb/day)* Ca (g/day)	P (g/day)	Mg (g/day)
50	110	9·6	6·2	0·5	18	12	0·8
100	220	15	7·3	1·2	27	13	1·7
200	441	18	9·8	4·0	30	15	5·0
300	662	21	15	5·5	33	20	6·5
400	882	26	24	7·0	37	29	8·0

Nearly all feeding trials, in which various levels of calcium intake have been given, have been made with cattle over 160 kg (352 lb) live weight.[19] Daily intakes of 17 and 21 g calcium were found to be adequate for calves of 175 kg (385 lb) and 215 kg (473 lb) live weight, respectively,[20,21] whereas the very low value of 5 g calcium per day was certainly inadequate.[22] More recently an intake of 8 g calcium per kilogramme dry matter for ruminant calves of 70–100 kg (154–220 lb) live weight has been recommended to give good mineralisation of the skeleton.[11] The normal serum calcium value is from 9 to 12 mg per 100 ml.

PHOSPHORUS REQUIREMENT

The endogenous faecal excretion of phosphorus of the very young calf is probably quite low, about 4·3–5·6 mg/kg body weight at 27–45 kg (60–100 lb),[1,23] but values of 12·6–15·6 mg/kg have been reported for calves in the weight range 182–227 kg (400–500 lb).[24–26]

The true absorption of phosphorus, as for calcium, is lower from dry foods than from milk, but endogenous urinary losses of phosphorus are much greater than those of calcium, probably about 15 mg phosphorus per kilogramme body weight. The true absorption of phosphorus from milk is between 94 and 99%,[1,7,9,10,27] provided that the intake of phosphorus is not excessive, when the value may be as low as 82%. On the other hand, with dry foods, the true absorption is about 73–76%.[24–26] Owing to high endogenous urinary loss of phosphorus, the availability of phosphorus is less than the true absorption, being about 82–90% for milk.

The net requirement of phosphorus for weight gain, calculated from data collected at slaughter indicates a value of about 10 g phosphorus per kilogramme weight gain at 50 kg (110 lb) live weight falling to 5·4 g/kg at 400 kg (882 lb) live weight.[18]

The gross requirement of phosphorus is obtained from the summation of the maintenance requirement and the net requirement for weight gain divided by the availability as a coefficient. The recently recommended requirements[18] are given in Table 3.1.

In general, the results of feeding trials have shown intakes of below 8–9 g phosphorus per day to be inadequate for calves up to 250 kg (550 lb) body weight, when assessed on the basis of normal growth and feed utilisation and maintenance of normal inorganic plasma phosphorus levels.[21,26,28–30] A requirement of 5 g phosphorus per kilogramme dry matter intake has recently been recommended in France as adequate for good mineralisation of the skeleton in ruminant calves of 70–100 kg (154–200 lb) live weight.[11] Normal plasma inorganic phosphorus levels are between 4 and 8 mg/100 ml; they increase from about 4–4·5 mg/100 ml at birth to about 5·5 mg/100 ml at 3 weeks of age.[31]

MAGNESIUM

It has been known for a number of years that whole milk supplemented with iron, copper, manganese and vitamin D is inadequate as the sole diet of the calf. This inadequacy was found to be associated with a shortage of magnesium in the diet, resulting in hypomagnesaemia and tetany (for clinical symptoms see p. 139).

The condition usually occurs with calves growing at a rapid rate on very large amounts of milk or milk substitute and having no access to other supplementary feeds. Recent work has suggested that on a high level of liquid feed, access to fibrous matter may exacerbate the risk of hypomagnesaemia because of the increased saliva production resulting

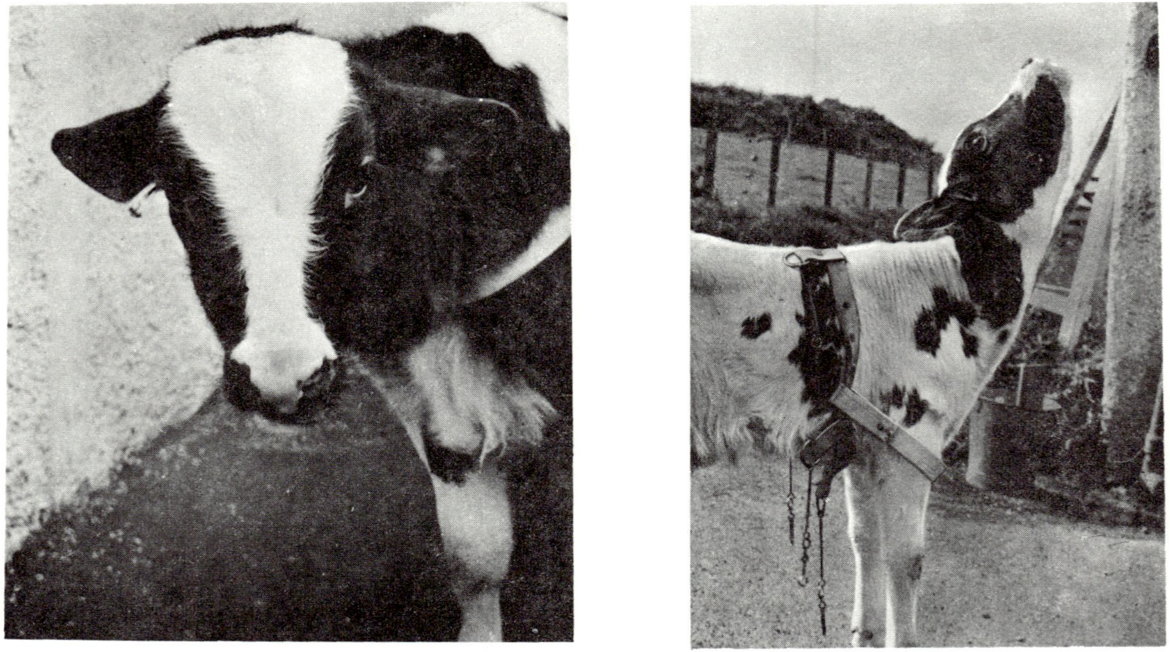

(a) (b)

Plate 3.1. Calf showing first signs of magnesium deficiency. (a) Change in the carriage of the head and ears, and (b) opisthotonus (arching of the spinal column so that the head is thrown backwards)[55]

in a greater loss of endogenous magnesium.[32,33] Moreover, hypomagnesaemia is considered to be an increasing problem in beef calves that are grazing during the 'spring flush'. The condition usually occurs during the first 3 months, during which the magnesium content of the blood may be steadily falling.

The normal plasma magnesium level of the calf is about 2·2–2·7 mg/100 ml and convulsions develop when the concentration has fallen to 0·3–0·7 mg/100 ml.[34] Even at these low plasma levels of magnesium, only slight clinical signs may occur at first, because depletion of magnesium from the bone is occurring. Normal calf bone contains about 0·7–0·8% magnesium, and convulsions are likely to occur when the concentration of magnesium in the bone has fallen to below 0·37%. Ratios of calcium to magnesium in the bones of affected animals are 90 : 1 or greater, compared with 60 : 1 or less in normal calves. As there appears to be little loss of magnesium from the soft tissues, which contain about 40% of the body magnesium, the total depletion of body magnesium is about 30%.[35]

There are a number of factors that may affect the incidence of hypomagnesaemia. A decline in availability of magnesium occurs with age, and is lower on dry than liquid diets. At 1 month of age, 75–90% of magnesium intake in milk was available,[8] but by 3 months of age the availability of magnesium in liquid diets had fallen to 40%.[36] This difference is related to the fact that, in the young calf, absorption takes place in both the large and small intestine, but in the older calf, possibly due either to physiological changes in the gut wall or to the development of the bacterial flora, absorption in the large intestine no longer occurs.[37,38]

Vitamin D has no effect on the absorption of magnesium,[39,40] but there is some evidence that excessive calcium in the diet aggravates the condition; addition of large amounts of calcium lactate or calcium carbonate to the diet of milk-fed calves has been shown to lower plasma and bone magnesium levels.[41] However, if calves have become hypocalcaemic due to a low vitamin D intake and are also suffering from hypomagnesaemia, supplements of magnesium cause a rise not only in plasma magnesium but also in plasma calcium.[42,43] Moreover, quite often in cases of hypomagnesaemia, the soft tissues become calcified,[44,45] possibly because with depletion of magnesium from the bone there is also a release of calcium.

Diarrhoea in calves will reduce absorption of magnesium, because of the more rapid rate of passage through the alimentary tract[46] and because of the higher excretion in the faeces of magnesium soaps of fatty acids. In the young calf with diarrhoea, the faecal output of magnesium was increased to 24 m-equiv. daily from 11 m-equiv. for the normal calf.[47]

The inclusion of fat of low digestibility in a milk substitute may also result in the increased formation of magnesium soaps,[48] which will render the magnesium no longer available to the calf.

There is some indication that low environmental temperatures may reduce the retention of magnesium from whole milk. At 4–9°C (39–48°F) magnesium retention at 2–11 weeks of age was 12% less than that of calves maintained at 20–32°C (63–90°F), although calcium retention was unaffected.[49] However, weight gains averaged 0·4 kg and 0·93 kg/day (0·9 and 2·1 lb/day) at the two temperatures, respectively, although digestibility of the dry matter was similar. Moreover, there appears to be an inverse relationship between thyroid activity and plasma magnesium levels. A low environmental temperature increases the concentration of protein-bound iodine in the blood, and this concentration has been shown to be inversely related to plasma magnesium.[50]

At a constant dry matter intake, magnesium is retained progressively less efficiently as the concentration in the diet increases.[35]

The ruminant calf, except when grazing, does not usually show signs of hypomagnesaemia because, although the availability of magnesium from dry food is lower than that from milk, the magnesium content of dry diets is usually about five times that of milk.

MAGNESIUM REQUIREMENT

The endogenous faecal magnesium of liquid-fed calves has been reported to be as low as 0·5 mg/kg body weight at 2–5 weeks of age,[36] 3–4 mg/kg body weight at 3–8 weeks[35] and 2·2 mg/kg for calves weighing only 128 kg (282 lb) at 26–32 weeks.[36] With diets low in magnesium the loss in the urine is negligible.

The availability of magnesium, as mentioned earlier is about 75–90% at one month of age,[8] falling to about 40% for milk-fed calves at 21–34 weeks.[36] With 6–9 month old steers, given hay or maize silage, the availability was estimated as 21–27%.[36]

The net requirement of magnesium for weight gain above maintenance is about 350–400 mg/kg.[51] An approximation of the magnesium retention can be obtained from calcium and nitrogen retention by the use of a ratio of Ca:Mg in bone of 45:1 and a ratio of N:Mg in soft tissue of 140:1.[35]

The recently recommended requirements for magnesium[18] are given in Table 3.1.

From a study of blood levels of magnesium and of magnesium retention, minimum requirements varying from 15 to 51 mg magnesium per kilogramme live weight have been suggested.[52–55]

With single-suckled calves, an intake of 1·6 g magnesium per day maintained normal serum magnesium levels, but 1·1 g/day was inadequate for this purpose.[56] When supplementing liquid diets, the laxative effect of magnesium must be considered. Inclusion of 0·88 g magnesium as magnesium oxide per kilogramme dry matter in a milk substitute diet for veal calves, obtaining about 1·3 g magnesium daily from the basic diet, caused a significant increase in serum magnesium at 12 weeks of age, but had no effect on weight gain. However, this quantity of magnesium significantly reduced the dry matter content of the faeces.[57] More recently, a lower supplementary level of 0·25 g magnesium as magnesium chloride per kilogramme dry matter added to the basic diet has been used successfully.

POTASSIUM

The requirement for potassium, which is present mainly in the soft tissues, is not known, but a deficiency is only likely to occur in preruminant calves that are suffering from diarrhoea.

The endogenous loss in the faeces of normal milk-fed calves is estimated as about 0·05–0·20 g/100 kg body weight, and in the urine as about 1 g/l. of urine.[18]

Normal milk-fed calves excrete about 4 g potassium daily in their urine,[47] but ruminant calves of 67–93 kg (148–205 lb) live weight excreted about 17 g potassium daily in their urine with a concentration of about 8 g potassium per litre.[58]

Calves at birth were found to contain about 1·97 g potassium per kilogramme body weight[51] and retention during growth has been estimated as 1·6 g potassium per kilogramme weight gain.[18]

In normal healthy calves, the availability of potassium is assumed to be 100%. For a 50 kg (110 lb) calf, on a milk substitute diet, an allowance of 2·6 g potassium has been suggested.[18] Calves kept for long periods on semi-purified diets containing about 0·11% potassium and having a daily intake of 1·5 g potassium showed marked changes in their electrocardiographs, whereas diets containing 0·35% potassium corresponding to an intake of about 4·4 g potassium per day gave normal results.[59]

Excessive potassium in the diet is toxic. A liquid diet containing 2·6% potassium in the dry matter caused severe muscular weakness, circulatory disturbances, and oedema of the limbs, followed usually by death.[60] Similarly, intravenous injections of potassium chloride into calves caused gradual cardiac arrest after plasma potassium levels had risen to 8 m-equiv./1.[61,62]

Calves with severe diarrhoea are usually in negative potassium balance;

such calves produce about 2 g potassium daily in their faeces, or as much as 6 g potassium in severe cases.[47] The depletion of potassium from the tissues of the calf is associated with high levels in the serum and this, together with depletion of potassium from the heart muscle, may be the cause of death of calves from an *Escherichia coli* localised intestinal infection (see p. 120).[63–65]

Elevated serum potassium levels have also been associated with the early stages of vitamin A deficiency in calves[66] and reduced serum potassium levels with the condition known as cerebrocortical necrosis[67] (see p. 85).

Normal serum potassium levels of young calves are between 5·0 and 5·5 m-equiv./1.

SODIUM AND CHLORINE

Unlike potassium, which is mainly an intracellular ion, sodium is mainly an extracellular ion. Thus the concentration of sodium is low in soft tissues but high in body fluids. Healthy, young pre-ruminant calves excrete about 1·5 g sodium daily in the urine and about 0·1 g daily in faeces.[47] On the other hand, ruminant calves of 67–93 kg (148–205 lb) excreted about 7 g sodium and 11 g chlorine daily in their urine at a concentration of 3 g/l. and 5 g/l., respectively.[58]

The estimated accretion of sodium and chlorine in the weight gain of calves has been assessed as 1·4 g sodium per kilogramme and 0·9 g chlorine per kilogramme weight gain, and the daily endogenous losses are considered to be 1·32 g sodium and 2·52 g chlorine per 100 kg body weight.[18] The recently recommended requirements[18] of sodium and chlorine, assuming an availability of 100%, are given in Table 3.2.

Table 3.2 DAILY REQUIREMENT FOR SODIUM AND CHLORINE (g/day) OF CALVES GAINING WEIGHT AT 0·5 kg (1·1 lb) AND 1·0 kg/day (2·2 lb/day)

Live weight		*Maintenance + 0·5 kg (1·1 lb) gain/day*		*Maintenance + 1·0 kg (2·2 lb) gain/day*	
(kg)	*(lb)*	*Na*	*Cl*	*Na*	*Cl*
50	110	1·6	1·8	2·3	2·2
100	220	2·4	3·1	3·1	3·5
200	441	4·1	5·7	4·8	6·1
300	662	5·8	8·3	6·5	8·7
400	882	7·5	10·9	8·2	11·3

In South Africa, 1·5 g sodium and 5 g chlorine daily were found to be adequate for normal growth of heifers, although the animals did not calve normally.[68] Moreover, an increased retention of calcium and phosphorus has been observed in calves of 68–92 kg (150–202 lb) live weight, given a hay of poor quality together with 11–16 g sodium chloride daily.[69]

In practical rations, the inclusion of 0·3–0·5% sodium chloride in the concentrate mixture is usually recommended or alternatively a salt lick can be provided.

Diarrhoea has a marked effect on the sodium balance of the pre-ruminant calf. The loss in the faeces may increase from 0·1 g/day up to 1 g/day or in severe cases up to 4 g/day.[47] Severe diarrhoea is reflected in a decline in serum sodium levels from normal values of 135–140 m-equiv./1. down to 126 m-equiv./1.[63,64]

REFERENCES

1. BLAXTER, K. L. and WOOD, W. A. *Br. J. Nutr.* **6**, 1 (1952)
2. COMAR, C. L., MONROE, R. A., VISEK, W. J. and HANSARD, S. L. *J. Nutr.* **50**, 459 (1953)
3. HANSARD, S. L., COMAR, C. L. and PLUMLEE, M. P. *J. Anim. Sci.* **13**, 25 (1954)
4. HANSARD, S. L., CROWDER, H. M. and LYKE, W. A. *J. Anim. Sci.* **16**, 437 (1957)
5. VISEK, W. J., MONROE, R. A., SWANSON, E. W. and COMAR, C. L. *J. Nutr.* **50**, 23 (1953)
6. LENGEMANN, F. W. *J. Dairy Sci.* **48**, 1718 (1965)
7. BLACKWOOD, J. H., MORRIS, S. and WRIGHT, N. C. *J. Dairy Res.* **7**, 228 (1936)
8. SMITH, R. H. *Biochem. J.* **67**, 472 (1957)
9. HUGHES, J. S. and CAVE, H. W. *J. Nutr.* **4**, 163 (1931)
10. DUNCAN, C. W. and HUFFMAN, C. F. *J. Dairy Sci.* **17**, 83 (1934)
11. GUEGUEN, L. and MATHIEU, C. M. *Annls Zootech.* **14**, 231 (1965)
12. ELLENBERGER, H. B., NEWLANDER, J. A. and JONES, C. H. *Bull. Vt agric. Exp. Stn* no. 558 (1950)
13. TROWBRIDGE, P. F., MOULTON, C. R. and HAIGH, L. D. *Res. Bull. Mo. agric. Exp. Stn* no. 30 (1919)
14. HAIGH, L. D., MOULTON, C. R. and TROWBRIDGE, P. F. *Res. Bull. Mo. agric. Exp. Stn* no. 38 (1920)
15. MOULTON, C. R., TROWBRIDGE, P. F. and HAIGH, L. D. *Res. Bull. Mo. agric. Exp. Stn* no. 55 (1922)
16. MITCHELL, H. H. and McCLURE, F. J. *Bull. natn. Res. Coun., Wash.* no. 99 (1937)
17. MITCHELL, H. H. *Comparative nutrition of man and domestic animals.* Vol. 1. (Academic Press, New York and London 1962)
18. Agricultural Research Council. *The Nutrient Requirements of Farm Livestock. No. 2 Ruminants. Technical Reviews.* (H.M.S.O. London 1965)
19. CONVERSE, H. T. *Tech. Bull. U.S. Dep. Agric.* no. 1092 (1954)
20. LINDSEY, J. B., ARCHIBALD, J. G. and NELSON, P. R. *J agric. Res.* **42**, 883 (1931)
21. ARCHIBALD, J. G. and BENNETT, E. *J. agric. Res.* **51**, 83 (1935)
22. OTTO, J. S. *Onderstepoort J. vet. Sci.* **10**, 281 (1938)
23. LOFGREEN, G. P. and KLEIBER, M. *J. Nutr.* **49**, 183 (1953)

24. TILLMAN, A. D. and BRETHOUR, J. R. *J. Anim. Sci.* **17**, 100 (1958)
25. TILLMAN, A. D. and BRETHOUR, J. R. *J. Anim. Sci.* **17**, 104 (1958)
26. TILLMAN, A. D., BRETHOUR J. R. and HANSARD, S. L. *J. Anim. Sci.* **18**, 249 (1959)
27. LOFGREEN, G. P., KLEIBER, M. and LUICK, J. R. *J. Nutr.* **47**, 571 (1952)
28. HUFFMAN, C. F., DUNCAN, C. W., ROBINSON, C. A. and LAMB, L. W. *Tech. Bull. Mich. agric. Exp. Stn* no. 134 (1933)
29. BEESON, W. M., BOLIN, D. W., HICKMAN, C. and JOHNSON, R. F. *Bull. Idaho agric. Exp. Stn* no. 240 (1941)
30. LONG, T. A., TILLMAN, A. D., NELSON, A. B., GALLUP, W. D. and DAVIS, B. *J. Anim. Sci.* **16**, 444 (1957)
31. HAWKINS, G. M., ROY, J. H. B., SHILLAM, K. W. G., GREATOREX, J. C. and INGRAM, P. L. *Br. J. Nutr.* **13**, 447 (1959)
32. SMITH, R. H. *J. agric. Sci., Camb.* **56**, 105 (1961)
33. SMITH, R. H. *J. agric. Sci., Camb.* **56**, 343 (1961)
34. SMITH, R. H. *Nord. Vet. Med.* **16**, Suppl. 1, 143 (1964)
35. BLAXTER, K. L. and ROOK, J. A. F. *J. comp. Path. Ther.* **64**, 176 (1954)
36. SMITH, R. H. *Biochem. J.* **71**, 306 (1959)
37. SMITH, R. H. *Nature, Lond.* **184**, 821 (1959)
38. SMITH, R. H. *Biochem. J.* **83**, 151 (1962)
39. HUFFMAN, C. F. and DUNCAN, C. W. *J. Dairy Sci.* **19**, 440 (1936)
40. SMITH, R. H. *Biochem. J.* **70**, 201 (1958)
41. ALLCROFT, R. and IVINS, L. N. *Nord. Vet. Med.* **16**, Suppl. 1, 217 (1964)
42. PARR, W. H. *Vet. Rec.* **69**, 71 (1957)
43. SMITH, R. H. *Nature, Lond.* **191**, 181 (1961)
44. MOORE, L. A., HALLMAN, E. T. and SHOLL, L. B. *Archs Path.* **26**, 820 (1938)
45. THOMAS, J. W. *Symposium on Magnesium and Agriculture* p. 131 (West Virginia University, 1959)
46. SMITH, R. H. *Nature, Lond.* **198**, 161 (1963)
47. BLAXTER, K. L. and WOOD, W. A. *Vet. Rec.* **65**, 889 (1953)
48. RAVEN, A. M. and ROBINSON, K. L. *Br. J. Nutr.* **12**, 469 (1958)
49. HOHNECK, E. B. *Tierärztliche Hochschule, Hannover: Inaug. Diss. Doct. Med. Vet.* (1959)
50. GIRARD, A., LARVOR, P. and BROCHART, M. *Annls Biol. anim. Biochim. Biophys.* **5**, 383 (1965)
51. HOGAN, A. G. and NIERMAN, J. L. *Res. Bull. Mo. agric. Exp. Stn* no. 107 (1927)
52. HUFFMAN, C. F., CONLEY, C. L., LIGHTFOOT, C. C. and DUNCAN C. W. *J. Nutr.* **22**, 609 (1941)
53. RAY, S. N. *Indian J. vet. Sci.* **12**, 204 (1942)
54. THOMAS, J. W. and OKAMOTO, M. *J. Dairy Sci.* **37**, 656 (1954)
55. BLAXTER, K. L., ROOK, J. A. F. and MacDONALD, A. M. *J. comp. Path. Ther.* **64**, 157 (1954)
56. BLAXTER, K. L. and SHARMAN, G. A. M. *Vet. Rec.* **67**, 108 (1955)
57. ROY, J. H. B. *Vet. Rec.* **76**, 511 (1964)
58. WOELFEL, C. G., CALHOUN, M. C., ROUSSEAU, J. E. Jr., EATON, H. D. and LUCAS, J. J. *Res. Rep. Univ. Conn. agric. Exp. Stn* no. 6 (1965)
59. SYKES, J. F. and ALFREDSON, B. V. *Proc. Soc. exp. Biol. Med.* **43**, 575 (1940)
60. BLAXTER, K. L., COWLISHAW, B. and ROOK, J. A. F. *Anim. Prod.* **2**, 1 (1960)
61. BERGMAN, E. N. and SELLERS, A. F. *Am. J. vet. Res.* **14**, 520 (1953)
62. BERGMAN, E. N. and SELLERS, A. F. *Am. J. vet. Res.* **15**, 25 (1954)
63. ROY, J. H. B., SHILLAM, K. W. G., HAWKINS, G. M., LANG, J. M. and INGRAM, P. L. *Br. J. Nutr.* **13**, 219 (1959)

64. FISHER, E. W. *Br. vet J.* **121**, 132 (1965)
65. FISHER, E. W. and McEWAN, A. D. *Br. vet. J.* **123**, 4 (1967)
66. EATON, H. D., ROUSSEAU, J. E. Jr., HALL, R. C. Jr., CALHOUN, M. C. and LUCAS, J. J. *Res. Rep. Univ. Conn. agric. Exp. Stn* no. 4 (1965)
67. CLEGG, F. G. *Vet. Rec.* **78**, 505 (1966)
68. THIELER, A., GREEN, H. H. and du TOIT, P. J. *J. agric. Sci Camb.* **17**, 291 (1927)
69. SHEEHY, E. J. and SENIOR, B. J. *J. Dep. Agric. Repub. Ire.* **34**, 1 (1936)

CHAPTER FOUR

The Nutrient Requirements of the Calf—Trace Elements

Deficiencies in certain trace elements occur in particular areas of the world. The trace elements known to be essential for cattle include copper, iron, manganese, cobalt, zinc, selenium and iodine.[1] In some cases the pasture is deficient in the particular mineral, whereas in others there is an adequate amount in the pasture but for some often unknown reason it cannot be utilised properly by the calf. This latter deficiency is known as a 'conditioned' deficiency.

COPPER

Copper deficiency tends to occur on fenland and sandy and peaty soils throughout Great Britain. It has been identified in such areas as Caithness,[2] Cheshire, Lancashire,[3] Shropshire, the Fens and the Thames Estuary.

Signs of copper deficiency after the first month of life are associated in Australia and New Zealand with diets containing less than 5 mg copper per kilogramme dry matter and with those which contain an excess of molybdenum or insufficient phosphorus.[4] In Great Britain, levels of less than 5 mg copper per kilogramme dry matter are rarely found in pasture. However, when grass is growing fast in the spring, its copper content tends to be at its lowest, whereas its molybdenum content is at its highest. Moreover, with diets containing high levels of inorganic sulphate, molybdenum, even at a low concentration, will interfere with the assimilation of copper.[5]

The symptoms of copper deficiency when calves are at pasture vary somewhat in different areas, but the dams are usually copper-deficient during pregnancy, and in beef herds the symptoms appear in the calves

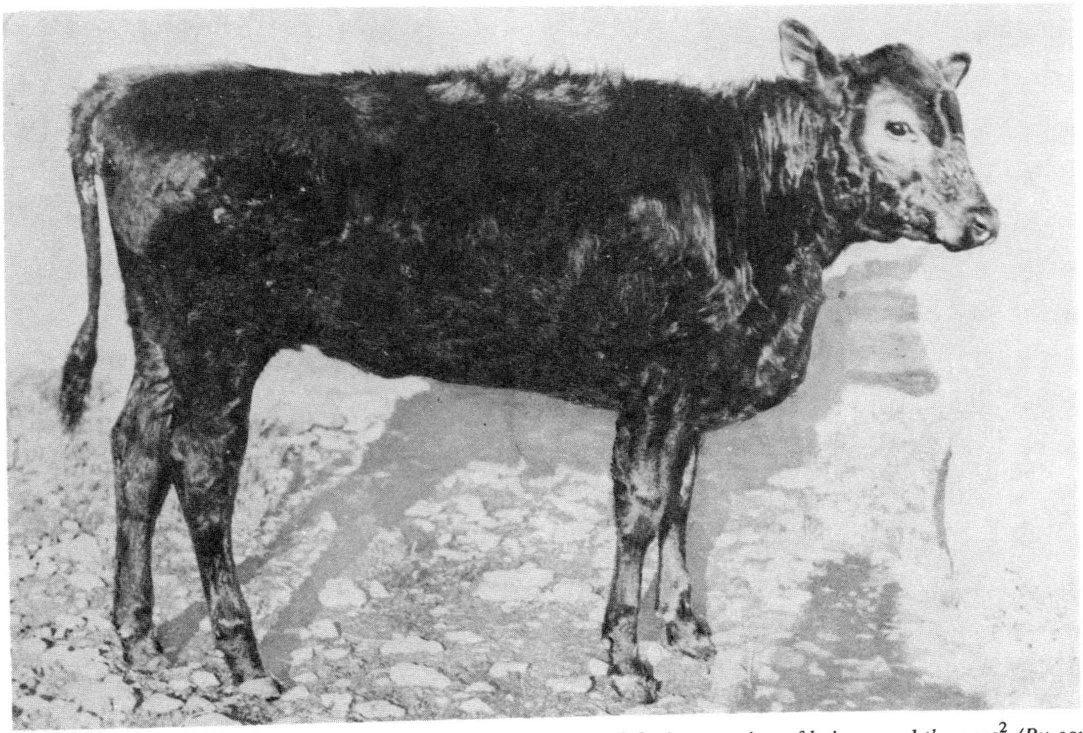

Plate 4.1. Copper deficiency in an Aberdeen-Angus calf showing pining, and depigmentation of hair around the eyes[2] (By courtesy of Cambridge University Press and by permission of the Controller of Her Majesty's Stationery Office. Crown Copyright)

at 2–3 months of age, or about 1 month after they are out at pasture. Dairy calves that have received concentrates do not usually show symptoms until they are out at pasture in their first or second year. Copper deficiency has also been reported in veal calves, and has been associated with the chewing of wooden veal crates, and a depression in growth rate after 70 days of age.[6]

The first sign of copper deficiency is a stilted gait, particularly of the hind legs, followed by a progressive loss of condition during the next 4 or 5 months, which may result in extreme emaciation and death. In black- or brown-coloured breeds the hair around the eyes becomes grey or dun coloured, and the pigmentation frequently extends down the forehead and lower jaw. This depigmentation occurs as a result of the failure of melanin formation, for which the copper-containing enzyme tyrosinase is required. In some cases, but not in all, diarrhoea occurs.

The copper content of normal new-born calves is about 14 mg/kg live weight[7] and the copper concentration in the liver of the new-born calf is about 4–8 times that of the dam.[8] Normal calves have a liver copper content of about 100 mg/kg dry matter, whereas deficient calves may have liver copper contents ranging from 2–16 mg/kg dry matter. Similarly, deficient calves have blood copper levels below 60 μg/100 ml.[9]

COPPER REQUIREMENT

Normal development of calves in copper-deficient areas has been obtained by administering copper. If it is suspected that calves are showing symptoms of copper deficiency, veterinary advice should be obtained. In the past, it was usually recommended that a drench containing 2 g copper sulphate should be given weekly, although some authorities advocated 1 g copper sulphate daily. The amount and frequency of dosage depends on the age of the animal and how severely the animal is depleted. Alternatively, prevention of copper-deficiency in calves has been achieved by drenching the dam with 10 g copper sulphate every 2 weeks during pregnancy. In these circumstances, there should be no need to supply copper to the calves during the subsequent grazing season.[10, 11] Spring-born calves from cows, given copper supplementation during the preceding winter, gained weight at 1 kg/day (2·2 lb/day) compared to 0·7 kg/day (1·5 lb/day) for calves of untreated cows.[12]

More recently, copper has been given in the form of copper-glycine by subcutaneous injection into the brisket. It has been recommended that an injection of 120 mg copper should be given to the dam in the last 2 months of pregnancy, followed thereafter by injection of the calf with the same amount at 4 months of age, and if necessary a further injection 4–6 months later.[9] 120 mg copper as copper glycine, given in a single injection to suckling calves at pasture, has been shown to increase weight gain during the summer on a farm where there was a previous history of copper deficiency.[13] Occasionally, large swellings occur after injection with copper glycine, reaching a maximum at 4–5 days of age and subsiding within 3–5 weeks after injection, but they apparently leave no after-effects.[9] Alternative methods that have been used for administering copper to calves include access to copper-rich mineral mixtures and the spraying of pastures with 6–11 kg copper sulphate per hectare (5–10 lb copper sulphate/acre). This latter method is effective when the pasture is deficient in copper, but not always satisfactory if the copper deficiency is a 'conditioned' one. Before administering copper to calves, it is essential that copper deficiency should be confirmed, owing to the risk of poisoning. If spraying of pastures is adopted, care must be taken if sheep also graze the land, because of their greater susceptibility to copper poisoning. Copper poisoning has also occurred in calves given milk substitutes containing 115 mg copper per kilogramme dry matter.[14]

With intensively reared pre-ruminant calves in Germany, daily intakes of 2·2 mg copper resulted in signs of copper deficiency after 70 days of age, the liver reserves at 140 days of age being 17·6 mg/kg dry matter. A daily intake of 7·7 mg copper was considered sufficient to meet the requirement of the calf up to this age. A reduced copper content of the kidney, blood and hair was found at the lower intake of copper, but no difference in the copper content of the muscle.[6] In the U.S.A., with calves given whole milk diets, a supplement of 6 mg copper daily maintained normal serum values, but had no effect on the haemoglobin levels of anaemic and normal calves,[15] in spite of the known interaction between copper and iron in haemoglobin synthesis in other species.

In the U.S.S.R., an optimum dietary level for 4–6 month old calves of 23 mg copper per kilogramme dry matter has been suggested.[16] In Great Britain the recently suggested dietary concentration for cattle is 10 mg/kg dry matter intake.[17]

IRON

Calves restricted to a diet of whole milk, or milk substitute based on animal or vegetable fats and skim milk, unsupplemented with iron, develop a microcytic (decrease in red-cell size) normochromic (red cells of normal haemoglobin content) anaemia, which, if allowed to continue, may develop into the hypochromic form (red cells of reduced haemoglobin content).[18,19]

According to one authority, pathological anaemia should be applied to haemoglobin values below 8 g/100 ml, and anaemia can be classified as follows: mild anaemia 7 g/100 ml, moderate anaemia 5–6 g/100 ml, severe anaemia 4 g/100 ml or less.[20,21] Calves having haemoglobin levels below 7 g/100 ml show clinical signs of anaemia, in particular anorexia (loss of appetite) followed by reduced digestibility of all nutrients, reduced efficiency of food conversion, pallor of the mucous membranes, enlarged livers and probably a greater susceptibility to diarrhoea and infections.

The age at which a condition of severe anaemia develops is affected by the rate of growth; the faster the growth rate the earlier the age. For instance, with Holstein calves gaining weight at only 0·68 kg/day (1·5 lb/day) and having initial haemoglobin values of 11·5 g/100 ml, haemoglobin levels did not fall to 5 g/100 ml until the calves were at least 289 days of age.[22] On the other hand, on very high levels of liquid feeding for veal production, the beneficial effects of supplementary iron are usually observed after 6–8 weeks of age,[18,19] but a response has been found during the first 2 weeks of life.[19] Indeed, calves may be anaemic at birth; 30% of those born at Ohio Experimental Station had haemoglobin levels of less than 9 g/100 ml at birth.[23]

Even in calves having access to dry food, blood haemoglobin concentration falls during the first 3–8 weeks of life and then recovers after weaning.[24–27] Similarly a fall in serum iron levels from 140–160 μg/100 ml at birth to 40–60 μg/100 ml between 10 and 25 days of age, with the values rising to 200 μg/100 ml after weaning has been reported,[28] although this finding has not been confirmed.[29] In beef calves suckling their dams in the North of Scotland, a mild anaemia (a 10% depression in haemoglobin level) was found between 40 and 80 days of life; administration of 35 mg iron per day as ferrous succinate increased haemoglobin levels and mean corpuscular volume to normal on most farms.[18] Under range conditions in the U.S.A., the prevention of even a slight fall in haemoglobin concentration from 13 g/100 ml at birth to 12 g/100 ml at 4 weeks of age, by intramuscular injection of 26·4 mg iron as iron-dextran per kilogramme body weight, was claimed to be associated with increased weight gains during this period.[30] With certain soil types, e.g. in Florida, forages will not maintain haemoglobin levels, and milk yields are reduced in these areas, especially on newly developed pasture, unless the cows receive a supplement of iron.[31] A different form of anaemia which is haemolytic in nature, resulting in a macrocytic (red cells of increased size), normochromic or hypochromic anaemia, with increased fragility of the red cells, has been reported in 12-month old steers fed solely on kale.[32]

For the production of veal calves, the irrational consumer-demand for white meat has resulted in the use of milk substitutes containing no supplementary iron. Where such milk substitutes are reconstituted with water of low iron content, calves are likely to become sufficiently anaemic to reduce their performance. For instance in the Netherlands, one of the leading countries in the production of veal, the water contains from 0·5 to 1 mg iron per litre or even as high as 17 mg/1., whereas in Southern England, water supplies were found to contain only traces of iron, the highest amount being 0·02 mg/1.; in these circumstances iron supplementation is likely to be necessary.[33] As a result of the Brambell Committee Report recommendations,[34] statutory regulations for the minimum content of iron in milk substitutes to prevent calves suffering from anaemia are likely to be introduced.

Sex and breed also influence blood haemoglobin levels, females having a higher content than males.[25] In an area around Shinfield, haemoglobin levels at birth of Ayrshire calves (12·1 g/100 ml) were significantly lower than those of Jersey and Friesian calves (13·0 and 13·3 g/100 ml, respectively). Moreover, the smaller the calf within a breed, the lower was the haemoglobin level at birth.[35] Similarly, in the Netherlands the red and white Rhine Meuse Yssel breed is reported to have haemoglobin levels 1·5 g/100 ml lower than the Friesian.[36] Differences between farms in the mean level of haemoglobin of calves at birth have also been noted. Low environmental temperatures may also affect haemoglobin levels, these being greater under cold conditions.[37]

With *ad lib.* liquid feeding of whole milk, a daily supplement of 240 mg iron and 7 mg copper had no effect on the performance of the pre-ruminant calf up to 39 days of age, at which time the unsupplemented calves had a haemoglobin level of 8 g/100 ml.[38] However, with Ayrshire calves given whole milk virtually *ad lib.*, the injection of 1·5 g iron as iron-dextran during the first 3 weeks of life significantly increased weight gain to 12 weeks of age, the effect being most marked from 8 weeks of age onwards. At 12 weeks of age the haemoglobin levels of the iron-supplemented calves were 7 g/100 ml compared to 4·5 g/100 ml for the control animals.[19] This level of iron supplementation increased liver reserves of iron from 33 mg to 50 mg but had no significant effect on muscle iron. However, when calves were injected with 500 mg iron at 10 day intervals, there was a marked effect on the colour of the meat, the greatest effect being on those with a high packed cell volume (haematocrit) at birth and resulted in nearly a 100% increase in the total haem pigments in the muscle.[39] The haem pigments in the muscle are in the form of both haemoglobin and myoglobin, the proportion of haemoglobin varying from 20 to 40% of the total. Although there is some debate as to whether muscle myoblogin is depleted very much in anaemia, results from slaughterhouse data show clearly that the myoglobin content of the muscle of veal calves (1·2 mg/g muscle) is lower than that of the ruminant calf (2·2 mg/g muscle).[40]

A detailed study[19] of the haematology during the first 12 weeks of life of pre-ruminant Ayrshire calves given 1·5 g iron as iron-dextran during the first 3 weeks of life showed that such injections reduced the decline with age of packed cell volume, haemoglobin content, mean corpuscle volume, mean corpuscle haemoglobin concentration and total leucocyte count. Iron supplementation also reduced the increase with age of erythrocyte sedimentation rate, lymphocyte monocyte and platelet counts, increased the percentage of band neutrophils and eosinophils present, and enhanced the increase in eosinophils with age. The erythrocyte count declined with age, but was unaffected by iron supplementation. Values at birth for these factors and the rate of decline with age in calves given no iron supplement are given in Table 4.1.

Table 4.1. CHANGE WITH AGE IN THE HAEMATOLOGY OF AYRSHIRE CALVES GIVEN UNSUPPLEMENTED WHOLE MILK *AD LIB.*[19]

	Value at birth	*Change with age (per day)*
Packed cell volume (%)	42·2	−0·26
Haemoglobin (g/100 ml)	12·1	−0·08
Erythrocyte count (10^{-6}/mm^3)	7·4	−0·02
Mean corpuscular volume (μm^3)	57·3	−0·25
Mean corpuscular haemoglobin concentration (%)	28·7	−0·03
Erythrocyte sedimentation rate (mm/24 hr)	2·2	+0·05
Total leucocyte count (10^{-3}/mm^3)	7·4	−0·05
Adult neutrophils (% of total leucocytes)	60·4	−0·5
Lymphocytes (% of total leucocytes)	39·6	+0·5
Platelet count (10^{-3}/mm^3)	338	+0·6

IRON REQUIREMENT

The endogenous loss of iron in 16-week old calves was found to be 12 mg/day and the availablility of iron in milk was only 26%, giving a dietary requirement for maintenance of 46 mg/day.[41]

At least three other estimates have been made of the iron requirements of the pre-ruminant calf. First, the net requirement for maintaining normal haemoglobin levels and adequate liver reserves for calves gaining 1 kg/day (2·2 lb/day) has been estimated to be about 50–60 mg iron per day.[18] Secondly, a net requirement for the calf of 227 kg (500 lb) live weight has been estimated as 1·2 mg iron per day to maintain a blood haemoglobin level of 10 g/100 ml and a further 16 mg/day for a weight gain of 0·9 kg/day (2·0 lb/day).[15] Assuming an availability of 30% the gross requirement would thus be 100 mg and 56 mg iron per day for the two estimates, respectively. Thirdly, with iron given by injection, e.g.

iron-dextran, the availability was estimated as between 60 and 70% and the gross requirement was estimated as 36 mg/day to maintain blood haemoglobin at 10 g/100 ml; this is equivalent to 70 mg/day if given by mouth, representing a concentration of 116 mg iron/kg dry matter intake.[19] In Germany, no difference was observed in the iron content of liver, kidney and spleen in calves given liquid diets supplying 59 mg or 104 mg iron daily up to 140 days of age, although the iron content in muscle, hair, bone and blood was increased at the higher level of intake.[6] A dietary concentration of 200 mg iron per kilogramme dry matter has recently been suggested,[17] but a value of 100–120 mg would probably be adequate.

It must be remembered that the availability of iron depends on its state of depletion in the calf; the availability being higher in a depleted calf.[42] Little is known of the availability of iron from dry diets, but in Great Britain the suggested dietary concentration is 30 mg/kg dry matter.[17] In the U.S.S.R., an optimum level for 4–6 month old calves of 140 mg/kg dry matter has been proposed.[16]

For supplementation of diets, little difference has been found between ferric and ferrous supplements, and it appears that the solubility of the salt is of more importance. The availabilities of ferrous sulphate, ferrous carbonate and ferric chloride did not differ significantly, but were in decreasing order. Ferric oxide was much less available. However, ferrous sulphate and ferric chloride increased serum iron levels significantly more than did ferrous carbonate.[42]

Excess of calcium in the diet may impair the assimilation of iron and an excess of iron may reduce the assimilation of phosphorus. Moreover, the addition of 240 mg iron per day to a milk diet given *ad lib.* to calves appeared to have a depressing effect on the vitamin A content of their livers.[38]

MANGANESE

Diets low in manganese content have been associated with sterility and abortions in dairy cattle[43] and deformities in calves at birth.[44,45] The particular deformities at birth are weak pasterns, 'overknuckling', enlarged joints, stiffness, twisted legs and shorter and more brittle humeri. No difference in birth weight of such calves has been found,[45] although poorer growth rate has been reported.[44] 'Overknuckling' in new-born calves is quite common, and normally the condition disappears within a few days. To what extent this condition is therefore due to a sub-clinical manganese deficiency is not known.

Manganese deficiency has been observed on peaty and sandy soils in the Netherlands and could be prevented by a dietary supplement of 2 g

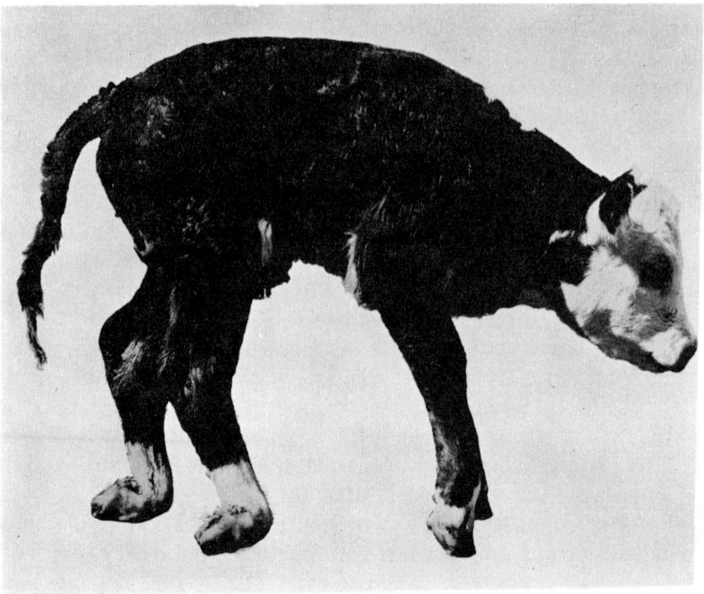

Plate 4.2. Overknuckling and deformation of the legs in a calf born to a manganese deficient cow[45]

manganese sulphate daily, or by treatment of the pasture with 15 kg manganese sulphate per hectare (13 lb/acre). It is probable that the manganese deficiency was a 'conditioned' one, as the pastures contained quite high levels of manganese (44–199 mg/kg). Manganese deficiency has also been reported in intensively reared pre-ruminant calves from 110 days of age onwards. The condition was associated with convulsive twitching of the tongue, together with swelling of the tarsal joints of the forelimbs and difficulty in walking.[6]

MANGANESE REQUIREMENT

Calves reared from 90 kg (220 lb) live weight on a ration containing less than 10 mg manganese per kilogramme basal diet were slightly slower to show signs of oestrus and were consistently slower to conceive than those given 40 mg manganese per kilogramme basal diet.[43] They also developed livers of abnormal structure, especially proliferation of the bile duct. The only organs studied that contained increased manganese after supplementation were the ovaries.

A more recent experiment compared cows given a diet containing 15·8 mg manganese per kilogramme dry matter with those given one containing 25·1 mg manganese per kilogramme dry matter. No clinical signs of manganese deficiency were apparent at the lower intake, but four services were required per conception compared with two on the higher intake; blood manganese was 1·8 and 2·8 μg/100 ml for the low and high manganese diets, respectively. However, all calves born from the dams on the low manganese diet were deformed at birth. The content of manganese in the blood of the calves was reduced from 2·1 to 1·9 μg/100 ml, in the bone from 1·5 to 1·0 mg/kg, in the liver from 11·8 to 6·9 mg/kg, in the kidneys from 2·5 to 1·1 mg/kg, and in the testicles from 2·7 to 1·7 mg/kg.[45]

The best indicator of manganese deficiency is considered to be the manganese content of the hair.[6,46] Calves from healthy adults are born with hair of low manganese content. In the Netherlands, the black hair of Friesian calves averaged 3 mg manganese per kilogramme at birth and increased to 7 mg manganese per kilogramme at 3 to 5 months of age, whereas adult animals had hair with an average content of 12 mg manganese per kilogramme.[46] In pre-ruminant calves showing deficiency symptoms after receiving 8·5 mg manganese daily, the manganese content of the hair at 140 days of age was 4·1 mg/kg compared with a content of 13·9 mg/kg in the hair of calves given 36 mg manganese daily. In this experiment no difference was established in blood, spleen or muscle manganese, but manganese in the liver was reduced from 12·5 to 8·6 mg/kg, in the kidneys from 4·3 to 2·8 mg/kg, and in bone from 3·2 to 1·4 mg/kg at the lower manganese intake.

In Great Britain, a dietary concentration of 40 mg manganese per kilogramme dry matter has recently been proposed as adequate.[17] Similarly, in the U.S.S.R. a value of 34 mg manganese per kilogramme is considered optimum for 4–6 month old calves.[16]

No detrimental effect was observed when a daily intake of 70 mg manganese per kilogramme basal diet was given,[43] but very high levels of 1 g/kg dry matter resulted in a slightly reduced weight gain and poorer feed conversion efficiency in 10-week old calves, and depressed blood haemoglobin.[47,48]

Excessively high intakes of calcium and phosphorus over a long period have been shown to reduce the weight gain of calves reared mainly on a milk diet, possibly due to interference with manganese absorption, since additional manganese (50 mg manganese per kilogramme basal diet) tended to counteract the excess.[49]

COBALT

Cobalt is necessary for the synthesis of vitamin B_{12} by rumen bacteria. A deficiency of cobalt thus causes a deficiency of vitamin B_{12}, a condition which prevents the animal from metabolising propionate.[50]

Cobalt deficiency occurs in many parts of the world, including Australia, Kenya, New Zealand, Scotland and in the U.S.A. The deficiency results in a decline in appetite, in severe emaciation and wasting of the musculature (marasmus or 'pining'), and in anaemia, which may be of the same type as in iron deficiency.[51]

A level of 40–60 μg cobalt per kilogramme in the livers of cattle indicate a cobalt deficiency compared with normal values of 80–120 μg/kg or more.[52] However, it is possible to maintain cattle on cobalt-deficient pastures by injection with vitamin B_{12}, without raising the liver cobalt content to normal. A better indicator of cobalt deficiency is therefore the vitamin B_{12} level of the liver, which in normal animals should contain more than 0·3 μg/g fresh liver to ensure optimum growth.[53]

In deficiency areas, the cobalt status of the pasture and soil is low (e.g. for herbage under 8 parts per 100 million). Calves born to cobalt-deficient cows are weak and do not generally survive for more than a few days.[54] Supplementation of the dam increases the cobalt content of the tissues of the calf, especially the liver and the kidneys. The cobalt content of new-born calves is from 0·1 to 0·2 mg cobalt per kilogramme dry matter, although skin and hair have a much higher level of 1 mg/kg dry matter. Supplementation of the dam increases the cobalt content of colostrum about 100-fold.

When a cobalt deficiency is diagnosed it can be overcome by feeding calves 3·5 ml (one teaspoonful ≡ 1/8 fluid ounce) of a cobalt sulphate solution (6 g/1. or 1 oz/gal) or, alternatively, by top-dressing the pastures with 2·2 kg/hectare (2 lb/acre) of cobalt chloride or sulphate mixed in with normal fertilisers. A more recent method of counteracting cobalt deficiency is the supplementation of cattle by the use of cobaltic oxide pellets of high specific gravity ('bullets') which usually remain in the reticulum.[53] If given in early pregnancy, the bullets have been shown to increase the liver storage of cobalt and vitamin B_{12} in the new-born calf. These bullets were found to be unsuitable for calves under 6 weeks of age, as the bullets were not satisfactorily retained.

COBALT REQUIREMENT

The requirement for cobalt of growing cattle is not known. However, as cobalt deficiency occurs in Australia on pastures containing about 40 μg cobalt per kilogramme dry matter, whereas cattle are normal on pastures containing 130 μg cobalt per kilogramme dry matter,[56] it has been suggested that the daily requirement is about 2·2 μg cobalt per kilogramme body weight.[57] The marked difference between this requirement and the toxic level of 1 mg cobalt per kilogramme body weight precludes the risk of poisoning under practical conditions.[58,59] However, a more recent experiment showed that with a basal ration estimated to supply 1·2–1·4 mg cobalt daily, a significant improvement in the weight gain of fattening heifers resulted from supplementation with a 20 g cobalt 'bullet'.[60]

In Great Britain, the suggested[17] dietary concentration for cobalt is 0·1 mg/kg dry matter, whereas in the U.S.S.R. a value of 1 mg/kg dry matter intake for the 4–6 month old calf is considered optimum.[16]

ZINC

Zinc deficiency in calves has been reported so far only with purified diets, although under grazing conditions it has been observed in yearlings and cows in Guyana, where the amount of zinc in the pasture was 18–42 mg/kg dry matter,[61] and also in housed cattle in Finland where the diet contained less than 50 mg zinc per kilogramme dry matter.[62]

Zinc is an essential component of the respiratory enzyme, carbonic anhydrase, and of a number of other enzymes and the highest concentrations are found in certain portions of the eye, the male sex glands and secretions, the hair and the bones.

A deficiency in calves is manifest first in a reduced growth rate;[63] this has been found to occur even at equal food intake, and digestibility of the diet does not seem to be affected.[64,65] Zinc deficiency also results in reduced blood haemoglobin and a reduced zinc content of most tissues other than muscle and brain.[66] In zinc deficient calves, faecal zinc excretion is reduced and urinary zinc excretion appears to be increased.[66,67] The gross clinical symptoms which follow appear in the order (*a*) excessive salivation, (*b*) loss of hair around the eyes and mouth, (*c*) hyperkeratosis of the skin of the neck, heel of the mandible and inner side of the legs, (*d*) enlargement of the hock bones, and (*e*) continual fidgeting with stepping movements of the hind legs.[68] Zinc deficiency has also been associated with reduced fertility of female cattle.[69]

If damage occurs to the skin of zinc-deficient calves in areas that do not normally develop parakeratosis, the area will develop the condition and is very slow to heal when the calf is subsequently given zinc; similar wounds in calves of normal zinc status heal much more rapidly. Moreover, the areas tend to produce grey hair in zinc-deficient animals.[70]

It has been suggested that a diagnosis of zinc deficiency in calves may be made if the plasma levels remain below 0·4 μg zinc per millilitre for at least 1 week.[68]

ZINC REQUIREMENT

Zinc deficiency has been produced in calves given diets containing 2·7 mg[63] and 3·6 mg/kg dry matter,[71] but no signs of deficiency

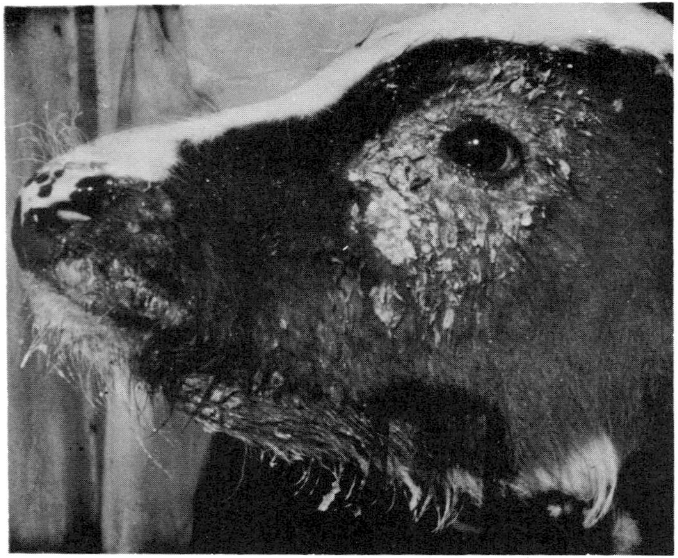

(a)

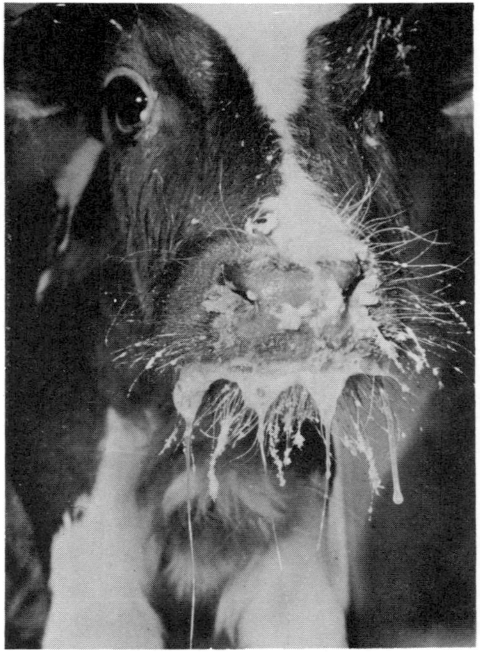

(b)

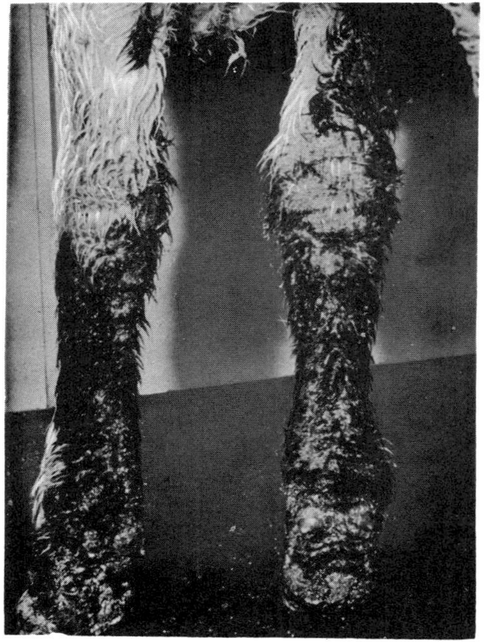

(c)

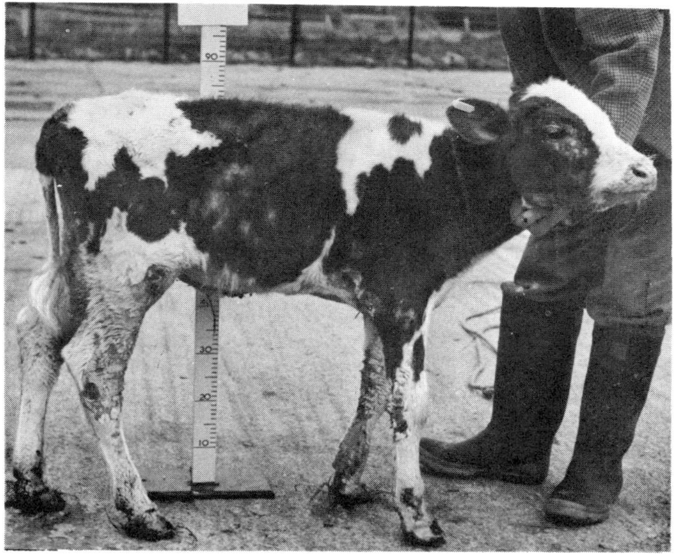

(d)

Plate 4.3. Zinc deficiency in calves. (a) Loss of hair and early skin lesions around the eye and mouth; (b) excessive salivation; (c) enlargement of hock bones, mild skin lesions, and changes in hoof structure; and (d) general appearance of a zinc-deficient Friesian calf. Note skin, leg and hoof lesions and staring coat[68] *(By courtesy of Cambridge University Press)*

occurred when the zinc content of the diet was increased to 46 mg/kg dry matter by the addition of zinc oxide. Further experiments have shown that an intake of 8·6 mg/kg dry matter should be adequate for normal growth,[72] whereas to maintain normal plasma levels a minimum intake of 10–14 mg/kg dry matter is required.[68]

Little is known of the availability of zinc, but the absorption of zinc from the alimentary tract of steers was found to be not greater than 50%.[73] Suggested dietary requirements of 50 mg/kg dry matter have been proposed in Great Britain[17] and of 42 mg/kg dry matter in the U.S.S.R.[16]

High levels of both calcium and magnesium in the diet have been implicated in increasing the zinc requirement.[62,69] Moreover, cadmium is an antimetabolite of zinc, and at high levels of intake is toxic; addition of zinc to the diet partially offsets the adverse effects of cadmium.[74]

IODINE

The requirement for iodine is mainly for the synthesis of the hormone thyroxine by the thyroid gland. Iodine deficiency is revealed in the calf by an enlargement of the thyroid (goitre) and a reduction in its iodine content. In the dam, plasma-bound iodine levels are below normal and abortions late in pregnancy or the birth of stillborn and weakly calves frequently occur.

Friesian cattle have lower plasma protein-bound iodine (2·4 μg/100 ml plasma) than Ayrshires (3·2 μg/100 ml), and Ayrshire cattle have lower values than Guernseys (4·0 μg/100 ml). This is presumably associated with differences in the basal metabolic rate of the three breeds (see p. 21) and also indicates that the Channel Island breeds may have a higher requirement of iodine. When suffering from iodine deficiency, the plasma protein-bound iodine of Friesians falls to 0·8–1·6 μg/100 ml plasma and of Guernseys to 1·6–3·2 μg/100 ml.[75]

The normal fresh weight of the thyroid gland of calves at birth is about 6·5–6·7 g[76,77] and at 3 weeks of age about 7·2 g[78] with no obvious breed or sex difference. Abnormally low thyroid weights have been found in a sandy coastal area of New Zealand.[76] Enlargement of the thyroid gland to 14·7–15·7 g at birth was found in calves born with goitre in Morayshire.[77] Similar high values of 12·5–14·0 g were found in a study of aborted foetuses, even though some showed degenerative changes and others did not. However, the iodine content of the thyroids showing degenerative changes was only 50 mg/100 g compared to 81 mg/100 g in the normal thyroids.[79] It is not known whether the degenerative changes in the thyroid are primarily the result of iodine deficiency, or are due to changes in the activity of the pituitary and adrenal glands.

There is some evidence that the iodine content of the thyroid of cattle is lower in the spring, but it is not usually possible to relate the condition to an iodine deficiency in the diet and the symptoms seem to be due to some interference with iodine metabolism in the physiology of the plants consumed by the animal, or in the physiology of the animal itself.

Areas of Great Britain associated particularly with the condition in cattle are also associated with thyroid enlargement in children, namely, North Oxfordshire, Berkshire, Buckinghamshire and Wiltshire. It has been suggested that the condition is associated with the hardness of the water in the area, but this has not been confirmed.[75] Iodine deficiency also occurs on sandy coastal soils in Morayshire, and in the mid-West and Western U.S.A.

The use of stabilised iodised salt licks, containing 0·0076% iodine, for pregnant cows may prevent the condition in calves, or alternatively, 1% of iodised salt may be included in the concentrate mixture. As goitrogens are found in leguminous crops, in particular white clover,[80] and also in kale[81] these may interfere with the conversion of inorganic to organic-bound iodine. The suggested dietary concentration of iodine is 0·12 mg/kg dry matter for calves and 0·8 mg/kg for pregnant and lactating animals in the absence of goitrogens in the diet, but 1·3 mg/kg and 2·0 mg/kg, respectively if goitrogens are present.[17]

Degenerative changes in the thyroid also occur in 'bulldog' calves of the Dexter and Friesian breeds.[82,83] The thyroid size of the calves at 3 weeks is also increased if the incidence of diarrhoea that has occurred prior to this time is high.[78]

REFERENCES

1. UNDERWOOD, E. J. *Trace Elements in Human and Animal Nutrition* (Academic Press, New York and London, 1962)
2. JAMIESON, S. and ALLCROFT, R. *Br. J. Nutr.* **4**, 16 (1950)
3. MORGAN, D. E. and CLEGG, A. *N.A.A.S. q. Rev.* **41**, 38 (1958)
4. NEAL, W. M., BECKER, R. B. and SHEALY, A. L. *Science, N. Y.* **74**, 418 (1931)
5. DICK, A. T. *Aust. vet. J.* **29**, 233 (1953)
6. ANKE, M. *Arch. Tierenähr.* **16**, 199 (1966)
7. RUSOFF, L. L. *Bull. Fla agric. Exp. Stn* no. 356 (1941)
8. COMAR, C. L., DAVIS, G. K. and SINGER, L. *J. biol. Chem.* **174**, 905 (1948)
9. ALLCROFT, R. and UVAROV, O. *Vet. Rec.* **71**, 797 (1959)
10. FIELD, H. I. *Vet. Rec.* **69**, 788 (1957)
11. FIELD, H. I. *Vet. Rec.* **69**, 832 (1957)
12. FIELD, H. I. *St. vet. J.* **8**, 1 (1953)
13. MORGAN, D. E., CLEGG, A., BROOKSBANK, N. H. and McCREA, C. T. *Anim. Prod.* **4**, 303 (1962)
14. SHAND, A. and LEWIS, G. *Vet. Rec.* **69**, 618 (1957)
15. MATRONE, G., CONLEY, C., WISE, G. H. and WAUGH, R. K. *J. Dairy Sci.* **40**, 1437 (1957).
16. CHUBINSKAYA, A. A. 'Mikroelementy v Zhivotnovodstvo. Min. sel.'-khoz.' *SSR. Vses. Akad. Sel.'-khoz. Nauk Otd. Zhivotnovodstvo* p. 107 (1962)

17. Agricultural Research Council. *Nutrient Requirements of Farm Livestock no. 2 Ruminants. Technical Reviews and summaries.* (H.M.S.O. London, 1965)
18. BLAXTER, K. L., SHARMAN, G. A. M. and MACDONALD, A. M. *Br. J. Nutr.* **11**, 234 (1957)
19. ROY, J. H. B., GASTON, H. J., SHILLAM, K. W. G., THOMPSON, S. Y., STOBO, I. J. F. and GREATOREX, J. C. *Br. J. Nutr.* **18**, 467 (1964)
20. HOLMAN, H. H. *Br. vet. J.* **111**, 440 (1955)
21. HOLMAN, H. H. *Br. vet. J.* **112**, 91 (1956)
22. HERMAN, H. A. *Res. Bull. Mo. agric. Exp. Stn* no. 245 (1936)
23. HIBBS, J. W., CONRAD, H. R. and GALE, C. *J. Dairy Sci.* **44**, 1184 (1961)
24. WISE, G. H., CALDWELL, M. J., PARRISH, D. B., FLIPSE, R. J. and HUGHES, J. S. *J. Dairy Sci.* **30**, 983 (1947)
25. THOMAS, J. W., OKAMOTO, M., JACOBSON, W. C. and MOORE, L. A., *J. Dairy Sci.* **37**, 805 (1954)
26. OWEN, F. G., VOELKER, H. H., JACOBSON, N. L. and ALLEN, R. S. *J. Dairy Sci.* 38, 891 (1955)
27. WING, J. M., JACOBSON, N. L. and ALLEN, R. S. *J. Dairy Sci.* **38**, 1006 (1955)
28. LANZ, H. *Schweizer Arch. Tierheilk.* **98**, 153 (1956)
29. BREMNER, K. C. *Aust. J. exp. Biol. med. Sci.* **44**, 259 (1966)
30. RICE, R. W., NELMS, G. E. and SCHOONOVER, C. O. *J. Anim. Sci.* **26**, 613 (1967)
31. WING, J. M. and AMMERMAN, C. B. *J. Anim. Sci.* **24**, 911 (1965)
32. GREENHALGH, J. F. D., SHARMAN, G. A. M., AITKEN, J. N. and DALGARNO, A. C. *Anim. Prod.* **9**, 286 (1967)
33. BAKKER, Tj. *Vet. Rec.* **72**, 702 (1960)
34. *Report of the Technical Committee to enquire into the Welfare of Animals kept under Intensive Husbandry Systems* (Cmnd. 2836). (H.M.S.O., London, 1965)
35. ROY, J. H. B., STOBO, I. J. F. and GASTON, H. J. *Rep. natn. Inst. Res. Dairy.* p. 52 (1966)
36. BAKKER, Tj. Personal communication (1966)
37. KAZAKOVA, E. M. *Trudȳ gos. Inst. ėksp. Vet.* **22**, 272 (1959)
38. NIEDERMEIER, R. P., ALLEN, N. N., LANCE, R. D., RUPNOW, E. H. and BRAY, R. W. *J. Anim. Sci.* **18**, 726 (1959)
39. CHARPENTIER, J. *Annls Zootech.* **13**, 165 (1964)
40. PATTERSON, D. S. P. and ALLEN, W. M. *Vet. Rec.* **78**, 823 (1966)
41. HANSARD, S. L. and FOOTE, L. E. *Fedn Proc. Fedn Am. Socs exp. Biol.* **20**, 296 (1961)
42. AMMERMAN, C. B., WING, J. M., DUNAVANT, B. G., ROBERTSON, W. K., FEASTER, J. P., and ARRINGTON, L. R. *J. Anim. Sci.* **26**, 404 (1967)
43. BENTLEY, O. G. and PHILLIPS, P. H. *J. Dairy Sci.* **34**, 396 (1951)
44. GRASHIUS, J., LEHR, J. J., BEUVERY, L. L. E. and BEUVERY-ASMAN, A. *Meded. Inst. Moderne Veevoeding 'de Schothorst'* 540 (1953)
45. ROJAS, M. A., DYER, I. A. and CASSATT, W. A. *J. Anim. Sci.* **24**, 664 (1965)
46. van KOETSVELD, E. E. *Tijdschr. Diergeneesk.* **83**, 229 (1958)
47. CUNNINGHAM, G. N., WISE, M. B. and BARRICK, E. R. *J. Anim. Sci.* **23**, 871 (1964)
48. CUNNINGHAM, G. N., WISE, M. B. and BARRICK, E. R. *J. Anim. Sci.* **25**, 532 (1966)
49. HAWKINS, G. E., WISE, G. H., MATRONE, G., WAUGH, R. K. and LOTT, W. L. *J. Dairy Sci.* **38**, 536 (1955)

50. MARSTON, H. R., ALLEN, S. H. and SMITH, R. M. *Nature, Lond.* 190, 1085 (1961)
51. FILMER, J. F. *Aust. vet. J.* **9**, 163 (1933)
52. McNAUGHT, K. J. *N.Z. Jl. Sci. Technol.* **30A**, 26 (1948)
53. SKERMAN, K. D., SUTHERLAND, A. K., O'HALLORAN, M. W., BOURKE, J. M. and MUNDAY, B. L. *Am. J. vet. Res.* **20**, 977 (1959)
54. MARSTON, H. R. *Physiol. Rev.* **32**, 66 (1952)
55. WARD, G. M., BENNE, E. J., WEBSTER, H. D., DUNCAN, C. W. and HUFFMAN, C. F. *J. Anim. Sci.* **8**, 632 (1949)
56. UNDERWOOD, E. J. and HARVEY, R. J. *Aust. vet. J.* **14**, 183 (1938)
57. FILMER, J. F. and UNDERWOOD, E. J. *Aust. vet. J.* **13**, 57 (1937)
58. ELY, R. E., DUNN, K.M. and HUFFMAN, C. F. *J. Anim. Sci.* **7**, 239 (1948)
59. KEENER, H. A., PERCIVAL, G. P., MORROW, K. S. and ELLIS, G. H. *J. Dairy Sci.* **32**, 527 (1949)
60. RICHARDSON, D., TSIEN, W. S., KOCH, B. A., WARD, J. K., SMITH, E. F. and BOREN, F. W. *J. Anim. Sci.* **19**, 1289 (1960)
61. LEGG, S. P. and SEARS, L. *Nature, Lond.* **186**, 1061 (1960)
62. HAARANEN, S. *Nord. VetMed.* **15**, 536 (1963)
63. MILLER, J. K. and MILLER, W. J. *J. Dairy Sci.* **43**, 1854 (1960)
64. MILLER, W. J., PITTS, W. J., CLIFTON, C. M. and MORTON, J. D. *J. Dairy Sci.* **48**, 1329 (1965)
65. MILLER, W. J., POWELL, G. W. and HIERS, J. M. Jr. *J. Dairy Sci.* **49**, 1012 (1966)
66. MILLER, W. J., BLACKMON, D. M., GENTRY, R. P., POWELL, G. W. and PERKINS, H. F. *J. Dairy Sci.* **49**, 1446 (1966)
67. MILLER, W. J., BLACKMON, D. M., POWELL, G. W., GENTRY, R. P., and HIERS, J. M. Jr. *J. Nutr.* **90**, 335 (1966)
68. MILLS, C. F., DALGARNO, A. C., WILLIAMS, R. B. and QUARTERMAN, J. *Br. J. Nutr.* **21**, 751 (1967)
69. HYPPOLA, K. *Maataloust. Aikakausk.* **38**, 180 (1966)
70. MILLER, W. J., MORTON, J. D., PITTS, W. J. and CLIFTON, C. M. *Proc. Soc. exp. Biol. Med.* **118**, 427 (1965)
71. MILLER, J. K. and MILLER, W. J. *J. Nutr.* **76**, 467 (1962)
72. MILLER, W. J., CLIFTON, C. M. and CAMERON, N. W. *J. Dairy Sci.* **46**, 715 (1963)
73. FEASTER, J. P., HANSARD, S. L., McCALL, J. T., SKIPPER, F. H. and DAVIS, G. K. *J. Anim. Sci.* **13**, 781 (1954)
74. POWELL, G. W., MILLER, W. J., MORTON, J. D. and CLIFTON, C. M. *J. Nutr.* **84**, 205 (1964)
75. ALLCROFT, R., SCARNELL, J. and HIGNETT, S. L. *Vet. Rec.* **66**, 367 (1954)
76. CAMPBELL, I. L., HOLLARD, M. G. and FLUX, D. S. *N.Z. Jl Sci. Techn* **31A**, no. 3, 29 (1949)
77. JAMIESON, S., SIMPSON, B. W. and RUSSELL, J. B. *Vet. Rec.* **57**, 429 (1945).
78. HAWKINS, G. M., ROY, J. H. B., SHILLAM, K. W. G., GREATOREX, J. C. and INGRAM, P. L. *Br. J. Nutr.* **13**, 447 (1959)
79. LOOSMORE, R. M., ALLCROFT, R., SALT, F. J., HUGHES, D. E., CHOU, J. T-Y. and WILSON, D. G. *Nature, Lond.* **193**, 595 (1962)
80. FLUX, D. S., BUTLER, G. W., JOHNSON, J. M., GLENDAY, A. C. and PETERSEN, G. B. *N.Z. Jl. Sci. Technol.* **38A**, 88 (1956)
81. SINCLAIR, D. P. and ANDREWS, E. D. *N.Z. vet. J.* **2**, 72 (1954)
82. SELIGMAN, C. G. *J. Path. Bact.* **9**, 311 (1904)
83. BERGER, J. and INNES, J. R. M. *Vet. Rec.* **60**, 57 (1948)

CHAPTER FIVE

The Nutrient Requirements of the Calf—Vitamins

The calf is born with little or no reserves of the fat-soluble vitamins A, D and E. The shortage of these vitamins at birth is overcome by the calf ingesting colostrum, which should contain a plentiful supply of carotene, the precursor of vitamin A, and the fat-soluble vitamins (see Volume 1, Table 3.1).

The concentration of these vitamins in colostrum is, however, affected by the feeding of the dam during pregnancy. In the winter, therefore, it is necessary to ensure that the dams receive a plentiful supply of green fodder, well-made hay or silage before they calve.

CAROTENOIDS AND VITAMIN A (RETINOL)

A deficiency of vitamin A in the dam, which is most likely to occur in intensively reared dairy heifers, dairy cattle housed throughout the year and beef cattle on poor winter rations, may result in abortion, a premature birth, or a calf born 'blind' or so weak that it is susceptible to infection. In the past, it was rarely necessary to supplement calf diets with vitamins under practical conditions, provided that the calves were supplied with high quality hay at an early age. However, with the development of intensive systems of rearing, which involve rapid growth rates and high intakes of concentrates or milk substitutes and correspondingly low intakes of hay, vitamin A supplements have become a necessity.

Artificially dried lucerne hay may contain as much as 243 mg carotene per kilogramme dry matter (111 mg/lb), whereas good quality hay and dried grass may contain up to 45 mg carotene per kilogramme (20 mg/lb); however, bleached meadow hays may contain less than 2 mg/kg (1 mg/lb). A considerable loss of carotene occurs during storage, the rate of loss being increased by a number of factors, including heat, light and oxygen.

Moreover, the rate of decline in carotene content has been shown to be related to the initial carotene content, the higher the carotene content when cured, the greater the rate of decline.[1] For instance, a lucerne hay with initial carotene content of 190 mg/kg dry matter (87 mg/lb) declined at a rate of 15 mg/kg (6·8 mg/lb) in each 28 day period, whereas a hay with an initial carotene content of only 15 mg/kg (7 mg/lb) declined at the much slower rate of 0·4 mg/kg dry matter (0·2 mg/lb) during the same period.

The pathological signs of acute vitamin A deficiency are manifold,[2, 3] but the young calf frequently dies as a result of pneumonia or severe diarrhoea.

The first detectable change with a marginal vitamin A deficiency is an increase in cerebro-spinal fluid pressure, and development of cysts in the pituitary. The increase in cerebro-spinal fluid pressure is probably responsible for the development of papilloedema (swelling of the optic disc where the optic nerve enters the retina of the eye). Spinal fluid pressures of over 160 mm saline are considered indicative of vitamin A deficiency, but a marginal deficiency may be occurring at values over 80 mm.[4] Moreover, normal calves tend to have a low pressure at birth, and this increases with age up to about 100 days.[5] Although it has been suggested that the increased cerebro-spinal fluid pressure increases the retinal venous pressure, and in turn the intraocular pressure, no difference in intraocular pressure between deficient and normal calves has been established.[6]

The eye lesions that follow result in night blindness owing to a shortage of visual purple. Besides the apparent blurring of the optic disc

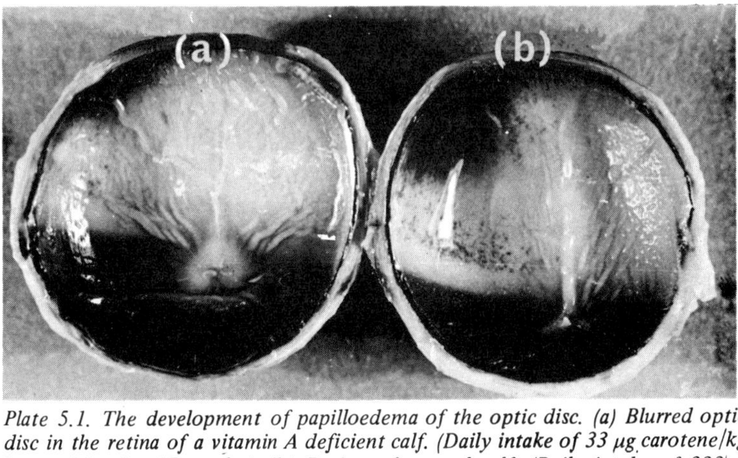

Plate 5.1. The development of papilloedema of the optic disc. (a) Blurred optic disc in the retina of a vitamin A deficient calf. (Daily intake of 33 μg carotene/kg live weight for 17 weeks.) (b) Retina of normal calf. (Daily intake of 333μg carotene/kg live weight to 17 weeks.) (MILLS, J. H. L., NIELSEN, S. W., ROUSEAU, J. E., WOEFEL, C. G. and EATON, H. D. Acta vet. Scand. 8, *324 (1967)*

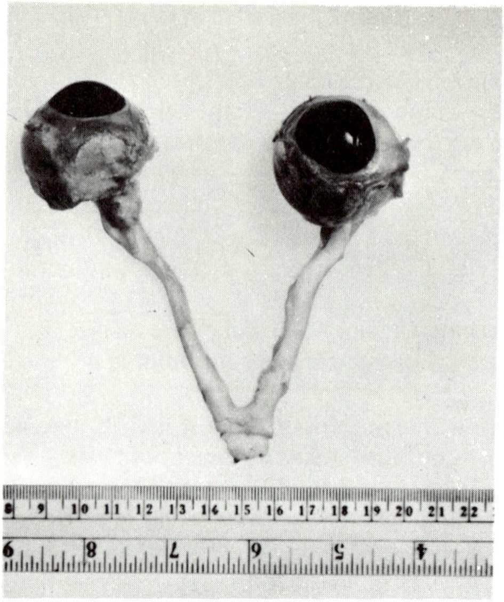

Plate 5.2(a). Eyes and optic nerve of normal calf[3]

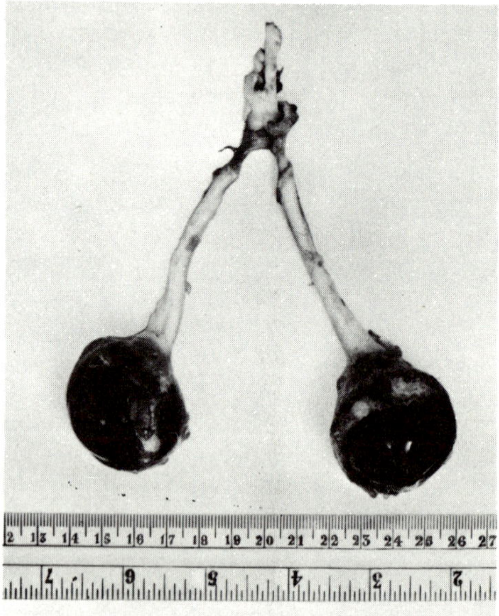

Plate 5.2(b). Eyes and optic nerve of vitamin A deficient calf. Note severe pinching of optic nerve anterior to optic chiasma[3]

Plate 5.2(c). Longitudinal section of optic nerve from (b) above. Note replacement of nerve by collagenous fibres at the point where nerve passed through the optic foramen[3]

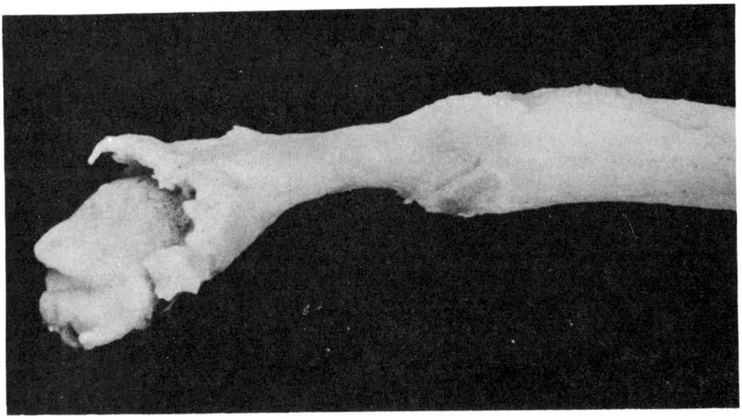

Plate 5.2(d). Vitamin A deficient calf. Constriction of the optic nerve as a result of vitamin A deficiency of a Friesian calf reared on a high-concentrate diet and barley straw ad lib. *(ABRAMS, J. T., BRIDGE, P. S., PALMER, A. C., SPRATLING, F. R. and SHARMAN, I. M.* Vet. Rec. **73**, *683 (1961))*

(Plate 5.1) the earliest symptoms that occur are bulging eyes with pupils dilated, owing to the failure of the irises to contract, fading of the tapetum lucidum and mottling of the tapetum nigrum. Finally, constriction of the optic nerve where it passes through the optic foramen results in atrophy of the optic nerve and complete blindness (Plate 5.2).[3,7-10] Constriction of the optic nerve is caused by the retarded growth of the membrane bone enclosing the central nervous system; this is also probably responsible for the increase in cerebro-spinal fluid pressure.[11,12] With mature animals that become vitamin A deficient, constriction of the optic nerve does not occur, since bone growth has been completed. Moreover, papilloedema does not occur so readily in mature animals as in young calves.[13]

In addition to these eye lesions, vitamin A maintains the integrity of the epithelial cells; a deficiency results in hyperkeratinisation (horny thickening) and lesions in the intestinal mucosa, which, by causing diarrhoea, may have a marked effect in depressing the utilisation of protein, and the digestion, absorption and metabolisability of energy.[14] Diarrhoea in calves, as a result of vitamin A deficiency, has frequently been reported.[15-20] Hyperkeratosis of the rumen (see p. 8) may also be associated with vitamin A deficiency. Metaplasma (transformation) of the epithelium may occur in the cervix, the mandilibular and paratid salivary glands, parotid gland ducts[2] and the bronchi.[2,3] An increased output of sodium and chloride in the urine, together with increased calcium deposits in the kidney[21] and increased plasma potassium levels[22] have also been reported.

Although diarrhoea in calves may occur when there is a vitamin A deficiency, the evidence shows that supplementary vitamin A has no discernible effect on the incidence of *Escherichia coli* infections (see p. 119), because protection against this condition is largely associated with the ingestion of antibodies from colostrum and with the feeding of milk substitute diets that do not predispose to the condition. Some years ago, it was considered[23-25] that colostrum-deprived calves could be successfully reared if given 7·5 mg vitamin A daily. However, it was shown later that an initial feed of 45 mg vitamin A[26] or 24 mg vitamin A, or 24 mg vitamin A together with 10 g lecithin had no protective value against *E. coli* infections.[27] Although calves are born with small liver reserves of vitamin A, no greater mortality was found during the first 3 weeks of life of calves born from normal dams and given a small quantity of separated colostrum, whether deprived of vitamin A from birth, or given 24 mg vitamin A in the first feed of life. Nevertheless, liver reserves were very low at 3 weeks of age, being 20 μg/100 g liver for the unsupplemented group and 30 μg/100 g liver for the supplemented group. The inclusion of lecithin with the vitamin A clearly increased the absorption of vitamin A so that liver reserves rose to 70 μg/100 g liver. It is known that calves with diarrhoea have lower reserves than normal, probably due to poor absorption[28] which in the early post-natal period can be related to the rate of excretion of the meconium (see Volume 1, p. 49).[27]

In spite of these findings in relation to *E. coli* infections, there is some evidence that weak calves born in the early spring months have low plasma carotene levels, and their dams have a low carotene content in both plasma and colostrum, due to a low carotene intake.[29]

Before considering the carotene and vitamin A requirement of the calf, some factors that influence this requirement must be considered. There is evidence that high energy and high protein intakes increase the requirement.[30] For instance, 12-week old calves gaining weight at 0·94 kg/day (2·08 lb/day) required 35% more carotene to maintain equivalent spinal fluid pressure and 16% more carotene to maintain equivalent vitamin A liver reserves per unit of live weight, than those gaining weight at 0·62 kg/day (1·36 lb/day).[31]

Evidence is accumulating that 'stress', whether induced by poor environmental conditions or by exposure to disease, may increase the vitamin A requirement.[32] For example, supplementation of liquid diets with the antibiotic chlortetracycline reduced the apparent requirement for vitamin A of veal calves up to the age of 13 weeks.[33] The minimum vitamin A requirement in winter is also possibly twice as high as in summer.[34,35]

The level of other fat-soluble vitamins in the diet may affect the utilisation of vitamin A. At very high levels of vitamin A intake (2 000 μg/kg live weight (1 000 μg/lb)), vitamin E supplementation caused an increased utilisation, but at very low levels (22 μg/kg (10 μg/lb)) of vitamin A intake, utilisation was depressed.[36] Calves that become rachitic, due to vitamin D deficiency, have lower levels of plasma vitamin A, although liver storage is unaffected,[37] whereas phosphorus-deficient cattle tend to have higher plasma carotene levels.[38]

There also appears to be a relationship between depressed thyroid function and a low vitamin A status, which is possibly due to a reduced rate of conversion of carotene to vitamin A in the intestinal wall.[39] A poorer conversion of carotene to vitamin A has also been suggested as the reason for the somewhat higher vitamin A requirement of the Channel Island breeds. A daily intake of 100 μg carotene/kg live weight (45 μg/lb) was found to be adequate for Holstein but not for Guernsey calves, which required 125 μg carotene per kilogramme (57 μg/lb).[40] Later work using spinal fluid pressure suggested that Guernsey and Jersey calves required 75 and 70 μg carotene per kilogramme live weight (34 and 32 μg/lb), respectively, these values being slightly greater than the requirement of 66 μg/kg (30 μg/lb) live weight found for Holsteins.[41]

CAROTENE AND VITAMIN A REQUIREMENT

Vitamin A requirements are often given in terms of international units; one international unit (i.u.) of vitamin A is equivalent to 0·3 μg vitamin A alcohol.

Except for the pre-ruminant calf, whose requirement for vitamin A is mainly obtained as such, requirements for vitamin A are best expressed in units of the precursor β-carotene, which is the normal source of vitamin A in ruminants and is converted into vitamin A in the walls of the small intestine. The requirement of β-carotene is about 4 to 5 times that of vitamin A itself,[40] but the relative efficiency depends on the method of testing the response[42] on the carrier used,[43,44] and also on the level at which the carotene is given. For instance, 3·8 μg carotene was needed to obtain the same response as 1 μg vitamin A at an intake of 88 μg carotene per kilogramme live weight (40 μg/lb), whereas 6·8 μg were required at an intake of 353 μg carotene per kilogramme live weight (160 μg/lb).[45] With ruminant calves, no difference has been found in the utilisation of vitamin A whether given in fish liver oil or in gelatin coated particles.[46] With pre-ruminant calves, addition of an emulsifying agent tends to increase absorption of both carotene and vitamin A.[27,47]

The requirement for vitamin A varies with the criteria used to assess adequacy; the most delicate test of incipient vitamin A deficiency is the failure to maintain a normal cerebro-spinal fluid pressure; changes in pressure will detect differences in intake of as little as 4·4 μg carotene per kilogramme live weight (2 μg/lb).[35]

REQUIREMENT OF THE DAM

The requirement of carotene during the last 3 months of pregnancy is considered to be about 3–5 times that for non-pregnant animals, although it has been suggested that a high intake is necessary throughout the gestation period.[48] A level of 178–264 μg β-carotene per kilogramme live weight (80–120 μg/lb) during the last 3 months of pregnancy resulted in normal calves, whereas 60 μg/kg (27 μg/lb) live weight was associated with the birth of weak calves.[49] In Great Britain, the recently suggested daily requirement of carotene for the pregnant heifer is 150 μg β-carotene per kilogramme live weight (68 μg/lb).[50] With intensive rearing of heifers indoors, in which little reserves of vitamin A have been built up from grazing during the summer, this recommendation may well be too low. In the U.S.S.R., an intake of carotene of 132–397 μg/kg (60–180 μg/lb) live weight during pregnancy was considered to be insufficient to prevent bronchopneumonia and gastrointestinal disorders in the calf, but 991–1 497 μg/kg (450–680 μg/lb) was adequate.[51] These abnormally high requirements may have been influenced by climatic conditions.

The main source of vitamin A for the young calf is from the ingestion

Table 5.1 EFFECT OF PREPARTUM DIET ON THE VITAMIN A AND CAROTENOID CONTENT OF COLOSTRUM AND POST-PARTUM BLOOD PLASMA OF THE DAM, AND ON THE VITAMIN A CONTENT OF THE PLASMA AND LIVER OF THE CALF (BEFORE SUCKING)[53]

Prepartum diet	*Dam*				*Calf*		*Colostrum and milk*					
	Plasma				*Plasma*	*Liver*						
	Vitamin A (μg/100 ml) Day		*Carotene (μg/100 ml) Day*		*Vitamin A (μg/100 ml)*	*Vitamin A (μg/100 g)*	*Vitamin A content (μg/g fat) Day*			*Carotene content (μg/g fat) Day*		
	1	7	1	7			1	3	7	1	3	7
Normal	19	16	390	286	4	75	43	15	5	34	10	3
Low carotene and vitamin A	12	10	60	57	2	21	47	17	5	11	6	2
High carotene	15	13	380	264	4	138	44	19	8	68	23	5
High vitamin A	24	22	272	187	8	4 940	174	48	13	30	12	3

of colostrum. The vitamin A and carotene content of colostrum can be increased by prepartum feeding of vitamin A and carotene, respectively. Moreover, such feeding increased the liver reserves in the calves at birth (see Table 5.1).[52,53]

REQUIREMENT OF THE CALF

In general, liver reserves of vitamin A are a better indicator of vitamin A intake than are plasma values. A great variation in reserves at birth is found under practical conditions; of 77 calves that had died before 4 weeks of age, 48% had livers with no demonstrable vitamin A, 15% had less than 300 μg/100 g liver and 37% had more than 300 μg/100 g liver [54]. Similarly, in the north of Scotland, 15% of livers had no detectable vitamin A, 37% had less than 600 μg/100 g, and 48% had more than 600 μg/100 g.[55] Since vitamin A given prepartum has a marked effect on liver reserves of the new-born calf which may be increased to as high as 4 300–5 600 μg/100 g of liver, and since the weight of the liver is about 17 g/kg body weight, this level of reserves might supply the vitamin A requirement of the calf for 5–6 weeks, assuming a requirement of 21 μg/kg (9·5 μg/lb) live weight.

Normal cerebro-spinal fluid pressure is maintained in Holstein calves, provided that the liver reserves of vitamin A are not less than 105 μg/100 g liver; at lower liver reserves, there is a linear increase in spinal fluid pressure, so that at a liver reserve of 10 μg/100 g liver, the estimated spinal fluid pressure would be 195 mm.[56] However, some calves may show papillaedema at higher liver reserves than 105 μg/100 g liver, if they have previously had plasma vitamin A levels below 10 μg/100 ml plasma.[57]

The maximum attainable liver reserve is not known, but liver reserves up to a mean of 74 200 μg/100 g was obtained in 9 animals, although the remainder of the animals in the weight range 69–268 kg (152–590 lb) did not exceed 7 900 μg/100 g.[58] Liver reserves of 1 460 μg/100 g liver were obtained in veal calves at 13 weeks of age having a daily intake of 44 μg vitamin A per kilogramme (20 μg/lb) from whole milk.[33]

Liver reserves can be related to plasma vitamin A levels by the equation[58]: $y=9{\cdot}08x-3{\cdot}7$ where y is the plasma vitamin A in μg/100 ml and x is the log liver concentration of vitamin A in μg/100 g.

At a liver reserve of 105 μg/100 g, it can be calculated that the plasma level is 14·7 μg/100 ml. This value can be compared with the findings that deficiency symptoms are likely to be found at plasma levels below 18 μg vitamin A per 100 ml.[59] A decline in appetite was found to occur when plasma vitamin A had fallen to 12–15 μg/100 ml[60] and night blindness has been associated with plasma levels below 10 μg/100 ml.[61]

Severe depletion of vitamin A is associated with plasma levels of 4–6 μg vitamin A per 100 ml.

Up to 1954, the requirements of vitamin A based on various criteria were as shown in Table 5.2.

Table 5.2 REQUIREMENTS OF VITAMIN A AND β-CAROTENE

	Vitamin A[62] (μg/kg live weight)	β-carotene[4] (μg/kg live weight)
Prevention of night blindness	8	24–35
Maximum growth rate	16	24–73
Prevention of squamous metaplasia of parotid gland duct	–	53
Prevention of papillaedema	–	53–79
Prevention of elevated spinal fluid pressure	–	66–73
Maximum plasma levels	128	
Maximum liver reserves	256	

More recent work has shown that there is elevation of spinal fluid pressure if carotene intake falls below 106 μg/kg live weight.[4] However, the carotene was given in the form of lucerne, which is known to have a lower utilisation than carotene in the form of oil. In Great Britain, the recently suggested daily requirement for calves is 80 μg β-carotene or 16 μg vitamin A per kilogramme live weight (36 μg/lb or 7 μg/lb).[50] From the recent findings given above, the recommendation may be somewhat low.

Too much vitamin A has a detrimental effect on the calf; some evidence indicates that as little as 138 μg vitamin A per kilogramme live weight (62·5 μg/lb) may reduce plasma ascorbic acid and tocopherol levels, whereas changes in the distribution of the protein in cerebro-spinal fluid occurred at intakes greater than 200 μg/kg (91 μg/lb) and changes in the mineralisation of the bones at greater than 1 327 μg/kg (603 μg/lb). At levels above 17 600 μg/kg live weight (8 000 μg/lb) gross clinical signs of hypervitaminosis A including hyperhidrosis (abnormal sweating), alopecia, hyperaemia (congestion of the mucosa) and lameness occurred.[63]

VITAMIN D

The calf requires vitamin D for growth and proper calcification of the bones; both exposure to sunlight and the feeding of good-quality hay will normally furnish the requirement.[64] The precursor ergosterol is converted to ergocalciferol (D_2) during the sun-curing of hay, whereas the precursor in the skin is converted to cholecalciferol (D_3) on exposure to sunlight. Both forms of vitamin D appear to be equally potent for the calf. In terms of roughage, the daily consumption of 1·3 kg lucerne hay per 100 kg body weight, or its equivalent as silage,

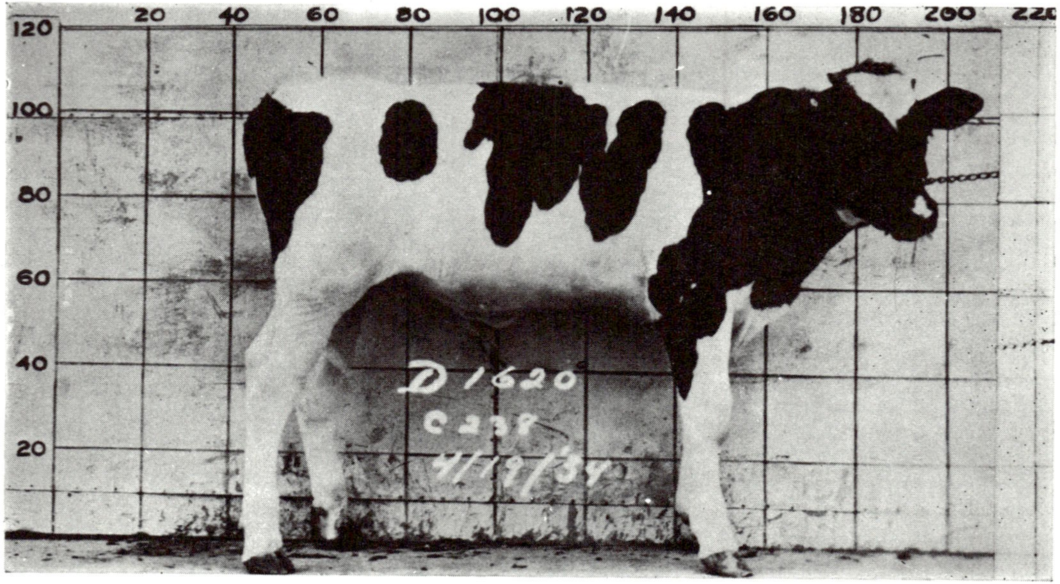

Plate 5.3. This calf had subclinical rickets. There were severe changes in the ribs and in the blood analysis, but the leg bones show no clinical evidence as yet. (After reference 70)

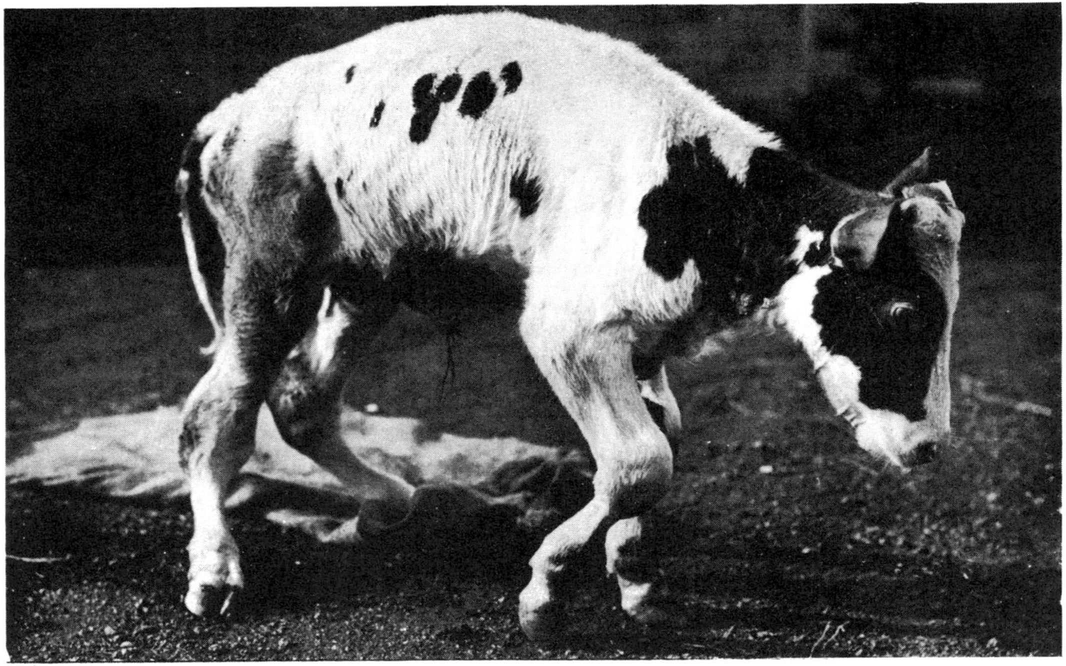

Plate 5.4. Calf suffering from severe rickets. Note emaciation, humping of back, swelling of joints, knuckling of pasterns and bowing of legs. (After reference 70)

will provide sufficient vitamin D if the calf has access to some sunshine.[65] However, the antirachitic power of the sun is very limited when the elevation of the sun is 35 degrees or less, which occurs in Britain during 5–6 months of the year.[66] A mature hay crop is likely to have a greater vitamin D content, but a lower carotene content. Even dried grass contains a moderate amount of vitamin D, as the grass has received considerable radiation before drying.

Vitamin D deficiency in the young animal causes rickets, even in the presence of adequate calcium and phosphorus.[65,67–69] The symptoms include a diminished appetite, the occurrence of digestive disturbance and a certain amount of tetany. These are followed by swelling and stiffness of the joints, a tendency for the long bones to be curved, resulting in a typical stilted walk, and a humped back.[70] (see Plates 5.3 and 5.4)

A shortage of phosphorus in the presence of adequate calcium and vitamin D will also give rise to rickets.[71] Even by increasing the calcium intake of calves on diets deficient in vitamin D, but adequate in phosphorus, the effect in reducing rickets is slight.[69] It has been suggested that magnesium has a sparing effect on vitamin D[72] and that, if large amounts of vitamin D are given to pre-ruminant calves before hypomagnesaemia (see p. 41) develops, the concomitant appearance of hypocalcaemia with the fall in blood magnesium will be prevented.[73]

A deficiency of vitamin D has also been shown to reduce the digestibility and retention of protein as well as of ash, and to increase basal metabolic rate.[74] High rates of weight gain may also increase the requirement of vitamin D per unit of live weight.

A deficiency of vitamin D in pregnant cows may result in bone abnormalities of calves at birth,[75,76] As with vitamin A, the vitamin D content of colostrum may be increased by prepartum feeding of vitamin D; by giving one million i.u. vitamin D daily to the dam for eight weeks before calving, the content of colostrum was increased three-fold, but the supplement did not raise plasma values or the liver reserves of the calves.[77]

VITAMIN D REQUIREMENT

An international unit of vitamin D is equivalent to the antirachitic activity of 0·025 μg crystalline D_3.

Early work with pre-ruminant calves suggested, on the basis of the composition of bone ash and blood calcium and phosphorus levels, that 0·8–1·0 i.u. vitamin D per kilogramme (0·36–0·45 i.u./lb) body weight was adequate for Holsteins.[78] Using similar criteria, together with X-ray analysis, the requirement of ruminant Holsteins was considered to be 6·6 i.u./kg (3 i.u./lb) body weight,[64] but with higher levels of phosphorus intake, a value of 3·9 i.u./kg (1·8 i.u./lb) body weight was deemed adequate for ruminant Holstein and Jersey calves, whereas 3·3 i.u./kg (1·5 i.u./lb) body weight was borderline.[65, 69, 79] More recently, a much higher requirement for pre-ruminant calves than for ruminant calves has been postulated, since although 4·5–7·0 i.u./kg (2·0–3·2 i.u./lb) body weight maintained normal serum calcium in 5–10 week old calves, calcium utilisation was considerably increased when vitamin D intake was raised to the high level of 900 i.u./kg (408 i.u./lb).[80] Moreover, an addition of 44 i.u./kg (20 i.u./lb) given daily to calves receiving a milk substitute already fortified with vitamin D, resulted in increased growth rate.[81] In the U.S.S.R., a supplement of 20–50 i.u. vitamin per kilogramme live weight (9–23 i.u./lb) in the form of irradiated fodder yeast resulted in a higher utilisation of nitrogen, calcium and phosphorus during the first 4 months of life than a supplement of 10 i.u./kg (4·5 i.u./lb) live weight.[82, 83]

For the Holstein heifer of 187 kg (412 lb) body weight, the lower value of 1·8–2·7 i.u./kg (0·8–1·2 i.u./lb) was adequate for normal growth,[84] a finding which suggests that the requirement may decrease with age.

In Great Britain the recently suggested daily requirement of vitamin D for calves is 4 i.u./kg (1·8 i.u./lb) live weight, for growing cattle 2·5 i.u./kg (1·1 i.u./lb) and for the pregnant cow 10 i.u./kg (4·5 i.u./lb) live weight.[50]

High levels of vitamin D are toxic for the young calf. Such toxicity has been produced giving 1 million i.u. vitamin D_3 daily.[85] As the daily dosage was increased from ½ to 4 million i.u. growth was progressively reduced, serum calcium values rose, those of magnesium fell and calcification of the vascular system became more pronounced.

VITAMIN E (TOCOPHEROL) AND SELENIUM

A deficiency of vitamin E results in muscular dystrophy, locally known by names such as 'white muscle disease', 'white flesh' and 'waxy muscle degeneration'. The symptoms are those of a much weakened musculature, but vary according to which muscles are affected. There is often a general instability of gait, a tendency for the hind legs to cross when the calf walks, 'winging' of the shoulder blades, straightening of the pastern joints and spreading of the claws of the feet, general weakness and inability to stand. Breathing, swallowing and the heart may also be affected. Death may be sudden, or the calf may linger on for some time. Wasting of the muscles does not, however, occur.

Vitamin E is present in considerable quantities in many dry foods, such as whole cereal grains and green fodder, and, in general, good quality hay should meet the vitamin E requirements of calves. Milk and animal products contain a low concentration of vitamin E and a supplement is normally necessary for milk substitutes fed at high levels for veal production, the amount being dependent on the quality of the fat. As with the fat soluble vitamins A and D, prepartum feeding of vitamin E increases the vitamin E content of colostrum.[86]

The trace element selenium is intimately linked with the function of vitamin E in protecting calves against muscular dystrophy, which can be of two forms.

ENZOOTIC MUSCULAR DYSTROPHY

It appears that enzootic muscular dystrophy is associated with areas of the world in which the soil content of selenium is low. The deficiency can be prevented either by the administration of small amounts of selenium or amounts of vitamin E that are large in relation to the normal requirement of the animal.[87]

For instance in Northern Ontario, the incidence of the disease is severe where the forage contains less than 1 mg selenium per kilogramme dry matter.[88] Similarly, in an area of north-east Scotland a high incidence of the condition in single-suckled calves was associated with dams given a diet consisting mainly of straw and turnips.[89] Calves born from out-wintered cows or from those given silage were rarely affected. Under conditions which resulted in inadequate amounts of vitamin E in colostrum, 150 mg α-tocopheryl acetate given daily to calves prevented the condition, whereas 10 mg was insufficient.[90] Further experiments showed that not less than 20 mg α-tocopheryl acetate daily would protect the majority of calves, whereas 0·25 mg selenium per day was slightly more effective.[91] In testing out other methods of selenium administration, it was found that a single subcutaneous injection of 15 mg selenium at birth or 5 mg at 3-weekly intervals from birth afforded considerable but not as complete protection as 0·25 mg selenium per day.[92]

In certain areas of the world, particularly in Nebraska, grazing animals suffer from selenium poisoning.[93] Owing to the high toxicity of selenium, prophylactic treatment should be given preferably in the form of vitamin E. By feeding about 2 mg tocopherol per kilogramme live weight prepartum to cows in north-east Scotland, enzootic muscular dystrophy was largely prevented.[94] Supplementation of the dam appears to be equally effective as supplementing the calf, although in Canada supplements of both selenium and vitamin E were superior to supplements of selenium alone.[88]

The level of serum glutamic oxaloacetic transaminase (SGOT) activity has been used as a criterion for diagnosis of muscular dystrophy.

In one experiment 31% of untreated calves had SGOT activity greater than 250 units/100 ml, compared with 2–5% of the selenium-treated calves.[92] Similarly, in Canada all sick calves had values greater than 1 100 units/ml but some calves from treated cows, although apparently healthy, also showed very high levels. In this latter experiment it was shown that there was no muscular dystrophy in calves from dams whose hair contained more than 0·25 mg selenium per kilogramme, whereas dams producing calves that developed muscular dystrophy, had a selenium content in the hair of between 0·06 and 0·23 mg/kg.[88]

MUSCULAR DYSTROPHY INDUCED BY DIETS CONTAINING UNSATURATED FATS

The inclusion of cod liver oil, lard and other fats, such as maize and linseed oil, containing a large proportion of polyunsaturated fatty acids in diets based on skim milk may cause muscular dystrophy.[95,96] Where a shortage of vitamins A or D in early life is suspected, these vitamins should be given in the form of a proprietary concentrate rather than in the form of fish liver oil. Although cod liver oil contains a considerable amount of vitamin E, it is insufficient to protect the calf against the effects of the unsaturated fatty acids present in the oil. 15–18 ml cod liver oil given daily was sufficient to overcome the protective action of as much as 50 mg DL α-tocopheryl acetate. Similarly, to prevent muscular dystrophy in calves given a tocopherol-deficient diet supplemented with 30 ml cod liver oil, 200 mg water dispersible D α-tocopheryl acetate were necessary.[97] Cod liver oil toxicity probably does not occur to the same extent in the ruminant calf, due to the hydrogenation of unsaturated fats in the rumen.[98] Cod liver oil when given to Guernsey calves as a supplement to whole milk resulted in a depletion of the body reserves of magnesium. Although this may have been due to the formation and excretion of magnesium soaps in the faeces (see p. 43), it was found that addition of either α-tocopherol or magnesium restored reserves to normal.[99]

Most other antioxidants, except methylene blue, seem to be ineffective in protecting calves. For instance, butylated hydroxyanisole (5 mg/kg dry matter) did not prevent muscular dystrophy in calves given skim milk containing 20% lard, or 10% lard and 10% beef dripping, unsupplemented with vitamin E, but no symptoms occurred with a similar diet containing 20% beef dripping.[100] The addition of 20 mg DL α-tocopheryl acetate per kilogramme dry matter, however, prevented the condition and maintained serum tocopherol values at the same level in calves given diets containing lard as in those containing beef dripping. Even so, the serum tocopherol levels resulting from these various diets may not have been really adequate for maximum performance, since a positive relationship was found between serum tocopherol level and weight gain, irrespective of the source of fat or of differences in initial live weight and food consumption.

VITAMIN E REQUIREMENT

An international unit of vitamin E is equivalent to the activity of 1 mg DL α-tocopheryl acetate.

For milk substitute diets containing about 20% fat the recommended inclusion rates are from 10 to 30 mg vitamin E per kilogramme (4·5–14 mg/lb) dry matter.[100–102] The higher value prevented muscular dystrophy in calves given a proprietary milk substitute containing 13% lard and 6·5% tallow.[102] In the absence of a source of polyunsaturated fatty acid, as little as 3 mg total tocopherol daily is sufficient to prevent the onset of muscular dystrophy in young calves.[96,103,104]

Although no precise requirements can be given for calves on diets low in selenium, it has been suggested[50] that the requirement of vitamin E under such conditions is from 0·7 to 0·9 mg/kg (0·3–0·4 mg/lb) live weight, compared with a requirement of 0·1–0·3 mg vitamin E per kilogramme live weight (0·05–0·14 mg/lb) for calves receiving normal diets.

VITAMIN B COMPLEX, VITAMIN C AND VITAMIN K

Although milk substitutes for calves are often supplemented with these vitamins, there is little evidence that the pre-ruminant calf is likely to suffer from a deficiency under practical conditions, except possibly of B_{12} (see below), although in Italy, vitamin B deficiency at 20 days of age in calves kept under practical conditions has been claimed.[105]

Vitamin C is synthesised in body tissues, but it has been suggested that supplementary vitamin C may prevent the infection of young calves through the navel.[24]

The synthesis of B-complex vitamins and vitamin K by the micro-organisms of the rumen make their inclusion in the diet unnecessary once the rumen is fully functional. Nevertheless, it would seem possible that the substrate may not always be ideal for synthesis of B vitamins during the first 2 months of life, if milk products have been completely replaced by dry food. Nevertheless, no difference in the content of B vitamins in the digestive tract of the calf over the range from 2 to 14 weeks, or even during the period from 8 to 32 days of age, has been found.[106,107] However, more recently a condition known as cerebro-cortical necrosis has occurred in ruminant calves usually between the ages of 3 and 6 months, a condition which can be cured by injection of thiamine (see below).

Pre-ruminant calves reared on purified diets have been shown to have a dietary need for riboflavine, thiamine, pyridoxine and vitamin B_{12}. Some indications of the content of B vitamins in the livers of new-born calves are given in Table 5.3.[108]

Table 5.3 B VITAMIN CONTENT OF LIVERS OF NEW-BORN CALVES (WEIGHT OF LIVER 816 g (720–997 g)

	μg/g liver	Range
Biotin	0·095	(0·08–0·12)
Folic acid	0·7	(0·3–0·9)
Nicotinic acid	74	(54–86)
Pantothenic acid	18	(16–22)
Riboflavine	9	(8–10)
Thiamine	1·2	(1·0–1·3)
Vitamin B_6 (pyridoxal)	1·9	(1·8–2·0)
Vitamin B_{12}	0·25	(0·15–0·36)

THIAMINE

Although it has been known for some years that the pre-ruminant calf has a requirement for thiamine[109,110] the symptoms after 7–10 days on the deficient diet, namely anorexia, diarrhoea and death from dehydration, did not appear to be very specific. Since then it has been shown that, after eliminating thiamine from the diet of pre-ruminant Holstein calves, deficiency symptoms occurred within 24–45 days after thiamine excretion in the urine had fallen below 0·05 mg/day. The deficiency symptoms were anorexia, heart arrhythmias, respiratory abnormality, excessive lachrymation and grinding of the teeth. Blood pyruvate and lactate increased from 1 and 15 mg/100 ml to 5 and 100 mg/100 ml, respectively and urinary pyruvate excretion increased from 15 mg/day to over 50 mg/day blood haemoglobin and haematocrit were depressed by 16 and 22%, respectively.[111]

The condition of cerebro-cortical necrosis occurs on a variety of diets, including pelleted grass meal, pasture with additional concentrates and also, apparently, on conventional hay and concentrate diets. Some of the symptoms are similar to those of hypomagnesaemia. The first signs are anorexia, followed by apparent blindness, staggering and tetanic spasms. Within 12–72 hr after onset of the symptoms the calf collapses, the ears droop and in the final stages the limbs and head are extended. General twitching of the musculature of the ears and eyelids, weaving of the head and neck and grinding of the teeth with groaning may occur; there also may be a brown diarrhoea. Death usually occurs within 3–6 days. The main lesions in such calves are necrotic areas in both cerebral hemispheres.[112,113]

Injections of thiamine appear to cure the condition if given at an early stage.[114] It appears that there is no failure in the synthesis of thiamine in the rumen in affected animals, and thus the condition must be associated with poor absorption of thiamine, or interference in its metabolism within the tissues.[115] It has been suggested that there is a similarity in the symptoms with those of salt poisoning; affected calves having lower serum potassium values (3·7 m-equiv./1.) than normal (4·8 m-equiv./1.).[116] More recently, it has been suggested that the condition may possibly be associated with the consumption of mouldy feeding stuffs.[117] One outbreak was associated with calves having access to mouldy bales of straw which were being used as a calf pen wall. The main fungi isolated was *Acrospeira macrosporoides* (Berk), which has been shown to produce thiaminase.[118]

RIBOFLAVINE

Symptoms of riboflavine deficiency have been produced either on 'synthetic' milk diets[119] or on 'photolysed' whole milk.[120] As with most B vitamin deficiencies symptoms of anorexia occur, coupled in the case of riboflavine with excessive lachrymation and salivation, and diarrhoea. Soreness of the corners of the mouth and shedding of the hair has also been associated with the deficiency. In deficient calves, urinary excretion of riboflavine was only 0·01 μg/ml compared to 1 μg/ml for a normal calf. Death occurred after about 6 weeks on the deficient diet.[121]

The requirement of riboflavine, using 'photolysed' milk diets was considered to be between 35 and 45 μg/kg (16 and 20 μg/lb) live weight.[120] Further work in the U.S.A., using purified diets, has suggested a requirement of 1 μg/g dry matter for Holstein calves, equivalent to 14 μg/kg (6 μg/lb) live weight,[123] whereas in Great Britain the requirement for Shorthorn and Friesian calves is considered to be not greater than 22 μg/kg (10 μg/lb) live weight.[124]

NICOTINIC ACID

Calves given liquid diets containing gelatin as a source of protein appeared to suffer from a deficiency of nicotinic acid.[125] The deficiency symptoms resemble those of black tongue in the dog. Even on diets devoid of nicotinic acid, no deficiency occurs providing proteins containing appreciable quantities of tryptophan, such as those in milk, are included.[126]

VITAMIN B_6 (PYRIDOXINE)

This vitamin has also been claimed as essential for the young calf.[126,127] A deficiency usually resulted in anorexia, diarrhoea and fits, with deaths occurring at 3–4 weeks of age, but some calves grew poorly to 4 months without showing definite symptoms. The post-mortem findings showed inflammation of the lymph nodes and some degree of enteritis.

PANTOTHENIC ACID

This vitamin has been claimed as essential for the calf.[126]

VITAMIN B_{12}

It has been suggested that there is a direct relationship between the intensity of rumen fermentation and the amount of B_{12} produced in the rumen, and that the concentration of B_{12} in the rumen declines with age.[128] A deficiency may occur in ruminant calves as a result of cobalt deficiency (see p. 59). In the pre-ruminant calf, oral doses of 0·20 and 0·28 μg/kg (0·09–0·13 μg/lb) live weight were considered to be barely sufficient to prevent vitamin B_{12} deficiency.[123] A requirement of 20–40 μg/kg (9–18 μg/lb) dry matter equivalent to about 0·5–1·0 μg/kg (0·23–0·45 μg/lb) live weight has been suggested,[129] whereas other work based on the results of single injections of B_{12} placed the requirement at 0·4 μg/kg (0·18 μg/lb) live weight.[130,131]

As the amount of vitamin B_{12} in milk is as low as 2·1 μg/kg,[132] it is possible that with calves restricted to liquid diets a deficiency might occur. No obvious signs of B_{12} deficiency were found in calves reared to 110 kg (243 lb) for veal production at an average weight of 1 kg/day (2·2 lb/day), when given an estimated daily intake of 0·54 μg B_{12}/kg (0·24 μg/lb) live weight.[33] As a precaution, it may be advisable to include in milk substitutes for veal calves a supplement of the order of 20 μg/kg (9 μg/lb) dry matter.

BIOTIN

Although it has been claimed in the U.S.A.[133] that a biotin deficiency state can be produced in calves, this has not been confirmed in Great

Britain, where attempts failed even when feeding raw egg white, as a source of avidin, or the biotin antagonist, desthiobiotin.[121] It was considered that the higher level of potassium and vitamin E used in the diets in Great Britain may have been responsible for this failure to confirm the American report.

REFERENCES

1. DOLGE, K. L., DEMBICZAK, C. M., ROUSSEAU, J. E. Jr., EATON, H. D., BEALL, G. and MOORE, L. A. *Bull. Univ. Conn. agric. Exp. Stn* no. 314 (1955)
2. HEMBOLDT, C. F., JUNGHERR, E. L., EATON, H. D. and MOORE, L. A. *Am. J. vet. Res.* **14**, 343 (1953)
3. LANGHAM, R. F. and HUFFMAN, C. F. *Mich. St. Coll. Vet.* **14**, 122 (1954)
4. EATON, H. D., ROUSSEAU, J. E. Jr., WOELFEL, C. G., CALHOUN, M. C., NIELSEN, S. W. and LUCAS, J. J. *Bull. Univ. Conn. agric. Exp. Stn* no. 383 (1964)
5. CALHOUN, M. C., ROUSSEAU, J. E. Jr., EATON, H. D., HALL, R. C. Jr. and LUCAS, J. J. *J. Dairy Sci.* **48**, 100 (1965)
6. WOELFEL, C. G., ROUSSEAU, J. E. Jr., KERSTING, E. J., NIELSEN, S. W. and LUCAS, J. J. *J. Dairy Sci.* **47**, 655 (1964)
7. MOORE, L. A., HUFFMAN, C. F. and DUNCAN, C. W. *J. Nutr.* **9**, 533 (1935)
8. McNUTT, S. H. and WALL, J. F. *Vet. Med.* **23**, 497 (1938)
9. MOORE, L. A. *J. Nutr.* **17**, 443 (1939)
10. WETZEL, J. O. and MOORE, L. A. *Am. J. Ophthal.* **23**, 499 (1940)
11. MOORE, L. A. and SYKES, J. F. *Am. J. Physiol.* **130**, 684 (1940)
12. MOORE, L. A. and SYKES, J. F. *Am. J. Physiol.* **134**, 436 (1941)
13. MOORE, L. A. *J. Dairy Sci.* **24**, 893 (1941)
14. RITZMAN, E. G., COLOVOS, N. F., KEENER, H. A. and TEERI, A. E. *Tech. Bull. New Hamps. agric. Exp. Stn* no. 87 (1945)
15. GUILBERT, H. R. and HART, G. H. *J. Nutr.* **10**, 409 (1935)
16. CUNNINGHAM, O. C. and ADDINGTON, L. H. *Bull. New Mex. agric. Exp. Stn* no. 257 (1938)
17. SCHMIDT, H. *Am. J. vet. Res.* **2**, 363 (1941)
18. HASTINGS, C. C. *J. Am. vet. med. Ass.* **100**, 190 (1942)
19. CREECH, G. T. and SEIBOLD, H. R. *Am. J. vet. Res.* **4**, 353 (1943)
20. MADSEN, L. L. and EARLE, I. P. *J. Nutr.* **34**, 603 (1947)
21. WOELFEL, C. G., CALHOUN, M. C., ROUSSEAU, J. E. Jr., EATON, H. D. and NIELSEN, S. W. *J. Dairy Sci.* **46**, 947 (1963)
22. EATON, H. D., ROUSSEAU, J. E. Jr., HALL, R. C. Jr., CALHOUN, M. C. and LUCAS, J. J. *Res. Rep. Univ. Conn. agric. Exp. Stn* no. 4 (1965)
23. LUNDQUIST, N. S. and PHILLIPS, P. H. *J. Dairy Sci.* **25**, 685 (1942)
24. LUNDQUIST, N. S. and PHILLIPS, P. H. *J. Dairy Sci.* **26**, 1023 (1943)
25. HANSEN, R. G., PHILLIPS, P. H. and RUPEL, I. W. *J. Dairy Sci.* **29**, 761 (1946)
26. ASCHAFFENBURG, R., BARTLETT, S., KON, S. K., TERRY, P., THOMPSON, S. Y., WALKER, D. M., BRIGGS, C., COTCHIN. E. and LOVELL, R. *Br. J. Nutr.* **3**, 187 (1949)
27. ASCHAFFENBURG, R., BARTLETT, S., KON, S. K., ROY, J. H. B., SEARS, H. J., THOMPSON, S. Y., INGRAM, P. L., LOVELL, R. and WOOD, P. C. *Br. J. Nutr.* **7**, 275 (1953)
28. SELLERS, K. C. and EDEN, E. *J. comp. Path. Ther.* **59**, 205 (1949)

29. KONERMANN, H. and EL FADLE, W. A. *Dt. tierärztl. Wschr.* **73**, 543 (1966)
30. ERWIN, E. S., GRAINGER, R. B., ALGEO, J. and GORDON, R. S. *J. Anim. Sci.* **20**, 931 (1961)
31. HAZZARD, D. G., GRIFO, A. P. Jr., ROUSSEAU, J. E. Jr., WOELFEL, C. G., EATON, H. D., NIELSON, S. W. and GOSSLEE, D. G. *J. Dairy Sci.* **45**, 91 (1962)
32. TILLMAN, A. D. *J. Am. vet. med. Ass.* **149**, 1571 (1966)
33. ROY, J. H. B., GASTON, H. J., SHILLAM, K. W. G., THOMPSON, S. Y., STOBO, I. J. F. and GREATOREX, J. C. *Br. J. Nutr.* **18**, 467 (1964)
34. KEENER, H. A., BECHDEL, S. I., GUERRANT, N. B. and THORPE, W. T. S. *J. Dairy Sci.* **25**, 571 (1942)
35. MOORE, L. A., BERRY, M. H. and SYKES, J. F. *J. Nutr.* **26**, 649 (1943)
36. DICKS, W. M., ROUSSEAU, J. E. Jr., EATON, H. D., TEICHMAN, R., GRIFO, A. P. Jr., and KEMMERER, H. A. Jr. *J. Dairy Sci.* **42**, 501 (1959)
37. THOMAS, J. W. and MOORE, L. A. *J. Dairy Sci.* **35**, 687 (1952)
38. ROSS, O. B. and GALLUP, W. D. *J. Anim. Sci.* **8**, 628 (1949)
39. CHANDA, R. and OWEN, E, C. *Biochem. J.* **51**, 404 (1952)
40. BOYER, P. D., PHILLIPS, P. H., LUNDQUIST, N. S., JENSEN, C. W. and RUPEL, I. W. *J. Dairy Sci.* **25**, 433 (1942)
41. MOORE, L. A., SYKES, J. F., JACOBSON, W. C. and WISEMAN, H. G. *J. Dairy Sci.* **31**, 533 (1948)
42. ROUSSEAU, J. E. Jr., EATON, H. D., HELMBOLDT, C. F., JUNGHERR, E. L., ROBRISH, S. A., BEALL, G. and MOORE, L. A. *J. Dairy Sci.* **37**, 889 (1954)
43. KON, S. K., McGILLIVRAY, W. A. and THOMPSON, S. Y. *Br. J. Nutr.* **9**, 244 (1955)
44. WISE, G. W., JACOBSON, N. L., ALLEN, R. S. and YANG, S. P. *J. Dairy Sci.* **41**, 143 (1958)
45. GRIFO, A. P., Jr., ROUSSEAU, J. E. Jr., EATON, H. D. and GOSSLEE, D. G. *J. Dairy Sci.* **43**, 1003 (1960)
46. GRIFO, A. P. Jr., ROUSSEAU, J. E. Jr., HAZZARD, D. G., WOELFEL, C. G., WALTER, J. C., EATON, H. D. and GOSSLEE, D. G. *Bull. Univ. Conn. agric. Exp. Stn* no. 365 (1961)
47. DVOŘÁK, M. *Vet. Med., Praha* **8**, 23 (1963)
48. RONNING, M., BEROUSEK, E. R., KUHLMAN, A. H. and GALLUP, W. D. *J. Dairy Sci.* **36**, 52 (1953)
49. CONVERSE, H. T. and MEIGS, E. B. *J. Dairy Sci.* **21**, A114 (1938)
50. Agricultural Research Council *Nutrient Requirements of Farm Livestock No. 2 Ruminants, Technical Reviews and Summaries* (H.M.S.O. London, 1965)
51. Anonymous *Veterinariya* **34**, 54 (1957)
52. SPIELMAN, A. A., THOMAS, J. W., LOOSLI, J. K., WHITING, F., NORTON, C. L. and TURK, K. L. *J. Dairy Sci.* **30**, 343 (1947)
53. WALKER, D. M., THOMPSON, S. Y., BARTLETT, S. and KON, S. K. *Int. Dairy Congr. XII, Stockholm* **1**, 83 (1949)
54. LOOSMORE, R. M. *Vet. Rec.* **76**, 1335 (1964)
55. North of Scotland College of Agriculture *Report on Investigations and Research,* **42** (1964)
56. EATON, H. R., ROUSSEAU, J. E. Jr., and LUCAS, J. J. *J. Dairy Sci.* **47**, 1016 (1964)
57. NIELSEN, S. W., MILLS, J. H. L., WOELFEL, C. G. and EATON, H. D. *Res. vet. Sci.* **7**, 143 (1966)
58. EATON, H. D., ROUSSEAU, J. E. Jr., DICKS, M. W., TEICHMAN, R. and GRIFO, A. P. Jr., *J. Dairy Sci.* **41**, 1456 (1958)

59. DAVIS, R. E. and MADSEN, L. L. *J. Nutr.* **21**, 135 (1941)
60. ABRAMS, J. T., BARNETT, K. C., BRIDGE, P. S., PALMER, A. C., SPRATLING, F. R. and SHARMAN, I. M. *Proc. Nutr. Soc.* **26**, i (1967)
61. GARTNER, R. J. W. and RYLEY, J. W. *Qd J. agric. Sci.* **19**, 363 (1962)
62. LEWIS, J. M. and WILSON, L. T. *J. Nutr.* **30**, 467 (1945)
63. HAZZARD, D. G., WOELFEL, C. G., CALHOUN, M. C., ROUSSEAU, J. E. Jr., EATON, H. D., NIELSEN, S. W., GREY, R. M. and LUCAS, J. J. *J. Dairy Sci.* **47**, 391 (1964)
64. BECHDEL, S. I., HILSTON, N. W., GUERRANT, N. B. and DUTCHER, R. S. *Bull. Pa agric. Exp. Stn* no. 364 (1938)
65. THOMAS, J. W. and MOORE, L. A. *J. Dairy Sci.* **34**, 916 (1951)
66. TISDALL, F. F. and BROWN, A. *J. Am. med. Ass.* **92**, 860 (1929)
67. RUPEL, I. W., BOHSTEDT, G. and HART, E. B. *Res. Bull. agric. Exp. Stn Univ. Wis.* no. 115 (1933)
68. GULLICKSON, T. W., PALMER, L. S. and BOYD, W. L. *Tech. Bull. Minn. agric. Exp. Stn* no. 105 (1935)
69. THOMAS, J.W. *J. Dairy Sci.* **35**, 1107 (1952)
70. BECHTEL, H. E., HALLMAN, E. T., HUFFMAN, C. F. and DUNCAN, C. W. *Tech. Bull. Mich. agric. Exp. Stn* no. 150 (1936)
71. THIELER, A. *Vet. J.* **90**, 143 (1934)
72. HUFFMAN, C. F. and DUNCAN, C. W. *J. Dairy Sci.* **18**, 605 (1935)
73. SMITH, R. H. *Biochem. J.* **67**, 472 (1957)
74. COLOVOS, N. F., KEENER, H. A., TEERI, A. E. and DAVIS, H. A. *J. Dairy Sci.* **34**, 735 (1951)
75. WALLIS, G. C. *J. Dairy Sci.* **21**, 315 (1938)
76. WALLIS. G. C. *Bull. S. Dak. agric. Exp. Stn* no. 372 (1944)
77. EATON, H. D., SPIELMAN, A. A., LOOSLI, J. K., THOMAS, J. W., NORTON, C. L. and TURK, K. L. *J. Dairy Sci.* **30**, 787 (1947)
78. HUFFMAN, C. F. and DUNCAN, C. W. *J. Dairy Sci.* **18**, 432 (1935)
79. MOORE, L. A., THOMAS, J. W., JACOBSON, W. C., MELIN, C. G. and SHEPHERD, J. B. *J. Dairy Sci.* **31**, 489 (1948)
80. SMITH, R. H. *Biochem. J.* **70**, 201 (1958)
81. KEYES, E. A. *Bull. Pa agric. Exp. Stn* no. 446. *Supl.* no. 2, p. 4 (1944)
82. SELJAMETOV, R. A. *Vest. sel'.-khoz. Nauki, Mosk.* No. 5, 87 (1963)
83. SELJAMETOV, R. A. *Zhivotnovodstvo, Mosk.* No. 2, 42 (1963)
84. BECHTEL, H. E., HUFFMAN, C. F., DUNCAN, C. W. and HOPPERT, C. A. *J. Dairy Sci.* **19**, 359 (1936)
85. BLACKBURN, P. S., BLAXTER, K. L. and CASTLE, E. J. *Proc. Nutr. Soc.* **16**, xvi (1957)
86. PARRISH, D. B., WISE, G. H., LATSCHAR, C. E. and HUGHES, J. S. *J. Nutr.* **40**, 193 (1950)
87. BLAXTER, K. L. *Proc. Nutr. Soc.* **21**, 211 (1962)
88. HIDIROGLOU, M., CARSON, R. B. and BROSSARD, G. A. *Can. J. Anim. Sci.* **45**, 197 (1965)
89. SHARMAN, G. A. M. *Vet. Rec.* **66**, 275 (1954)
90. BLAXTER, K. L. and McGILL, R. F. *Vet. Revs Annot.* **1**, 91 (1955)
91. SHARMAN, G. A. M., BLAXTER, K. L. and WILSON, R. S. *Vet. Rec.* **71**, 536 (1959)
92. BLAXTER, K. L., McCALLUM, E. S. R., WILSON, R. S., SHARMAN, G. A. M. and DONALD, L. G. *Proc. Nutr. Soc.* **20**, vi (1961)
93. FRANKE, K. W. and POTTER, van R. *J. Nutr.* **10**, 213 (1935)
94. BLAXTER, K. L. and SHARMAN, G. A. M. *Nature, Lond.* **172**, 1006 (1953)
95. BLAXTER, K. L. and BROWN, F. *Nutr. Abstr. Rev.* **22**, 1 (1952)
96. BLAXTER, K. L., WATTS, P. S. and WOOD, W. A. *Br. J. Nutr.* **6**, 125 (1952)

97. MAPLESDEN, D. C. and LOOSLI, J. K. *J. Dairy Sci.* **43**, 645 (1960)
98. HOFLUND, S., HOLMBERG, J. and SELLMANN, G. *Cornell Vet.* **46**, 51 (1956)
99. DEHORITY, B. A., ROUSSEAU, J. E. Jr., EATON, H. D., MYERS, G. S. Jr., GRIFO, A. P. Jr., DICKS, M. W., HAZZARD, D. G., HELMBOLDT, C. F., GOSSLEE, D. G., THOMAS, J. W., SYKES, J. F. and MOORE, L. A. *J. Dairy Sci.* **44**, 58 (1961)
100. ROY, J. H. B. *Vet. Rec.* **76**, 511 (1964)
101. FRENS, A. M. *Tijdschr. Diergeneesk.* **86**, 1636 (1961)
102. TOLLERSRUD, S. *Nord. VetMed.* **15**, 543 (1963)
103. BLAXTER, K. L., WOOD, W. A. and MACDONALD, A. M. *Br. J. Nutr.* **7**, 34 (1953)
104. BLAXTER, K. L., BROWN, F. and MACDONALD, A. M. *Br. J. Nutr.* **7**, 287 (1953)
105. ZACCHI, B. *Acta med. vet., Napoli* **8**, 25 (1962)
106. KESLER, E. M. and KNODT, C. B. *J. Dairy Sci.* **34**, 145 (1951)
107. KESLER, E. M. and KNODT, C. B. *Proc. Soc. exp. Biol. Med.* **79**, 85 (1952)
108. FORD, J. E. and GREGORY, M. E. *Rep. natn. Inst. Res. Dairy.* p. 123 (1964)
109. JOHNSON, B. C., HAMILTON, T. S., NEVENS, W. B. and BOLEY, L. E. *J. Nutr.* **35**, 137 (1948)
110. KON, S. K. and PORTER, J. W. G. *Rep. natn. Inst. Res. Dairy.* p. 78 (1952)
111. BENEVENGA, N. J., BALDWIN, R. L. and RONNING, M. *J. Nutr.* **90**, 131 (1966)
112. TERLECKI, S. and MARKSON, L. M. *Vet. Rec.* **73**, 23 (1961)
113. SPENCE, J. B., STEVENS, A. J., SAUNDERS, C. N. and HARRIS, A. H. *Vet. Rec.* **73**, 28 (1961)
114. DAVIS, E. T., PILL, A. H., COLLINGS, D. F., VENN, J. A. J. and BRIDGES, G. D. *Vet. Rec.* **77**, 290 (1965)
115. LEWIS, G., TERLECKI, S., MARKSON, L. M., ALLCROFT, R. and FORD, J. E. *Proc. Nutr. Soc.* **26**, xiii (1967)
116. CLEGG, F. G. *Vet. Rec.* **78**, 505 (1966)
117. EDWIN, E. E., LEWIS, G. and ALLCROFT, R. *Vet. Rec.* **83**, 176 (1968)
118. DAVIES, E. T., PILL, A. H. and AUSTWICK, P. K. C. *Vet. Rec.* **83**, 681 (1968)
119. WIESE, A. C., JOHNSON, B. C., MITCHELL, H. H. and NEVENS, W. B. *J. Nutr.* **33**, 267 (1947)
120. WARNER, R. G. and SUTTON, T. S. *J. Dairy Sci.* **31**, 976 (1948)
121. KON, S. K. and PORTER, J. W. G. *Rep. natn. Inst. Res. Dairy.* p. 83 (1951)
122. BRISSON, G. J. and SUTTON, T. S. *J. Dairy Sci.* **34**, 28 (1951)
123. DRAPER, H. H. and JOHNSON, B. C. *J. Nutr.* **46**, 37 (1952)
124. PORTER, J. W. G. *Rep. natn. Inst. Res. Dairy.* p. 88 (1953)
125. BLAXTER, K. L. and WOOD, W. A. *Br. J. Nutr.* **6**, 56 (1952)
126. JOHNSON, B. C., MITCHELL, H. H., HAMILTON, T. S. and NEVENS, W. B. *Fedn Proc. Fedn. Am. Socs exp. Biol.* **6**, 410 (1947)
127. KON, S. K. and PORTER, J. W. G. *Rep. natn. Inst. Res. Dairy.* p. 78 (1952)
128. PŠENIČNYJ, P. J. and ŠEVČENKOVA, D. I. *Sb. čsl. Akad. zeměd. Věd, Živoč. Výr.* **7**, 769 (1962)
129. LASSITER, C. A., WARD, G. M., HUFFMAN, C. F., DUNCAN, C. W. and WEBSTER, H. D. *J. Dairy Sci.* **36**, 997 (1953)
130. SMITH, S. E., KOCH, B. A. and TURK, K. L. *J. Nutr.* **44**, 455 (1951)
131. SMITH, S. E. and LOOSLI, J. K. *J. Dairy Sci.* **40**, 1215 (1957)
132. GREGORY, M. E., FORD, J. E. and KON, S. K. *J. Dairy Res.* **25**, 447 (1958)
133. WIESE, A. C., JOHNSON, B. C. and NEVENS, W. B. *Proc. Soc. exp. Biol. Med.* **63**, 521 (1946)

CHAPTER SIX

Antibiotics

The uses of antibiotics in calf rearing are two-fold; first for controlling potential pathogens, particularly in the pre-ruminant calf, and secondly for effecting an increase in food intake in the young ruminant calf. With ideal management and facilities, the routine use of antibiotics in calf-rearing should be unnecessary and should be restricted to those occasions when a life is endangered. These ideal conditions can be found on farms where calves are home-bred, colostrum intake is generous, the subsequent milk substitute diet is of good quality, the building is well-ventilated, the number of calves in a unit is small, and the calf rearer is experienced. Unfortunately, the existence of this combination of factors is exceptional. With calves that are not brought quickly and directly from the place of birth to the site of rearing, routine antibiotic feeding may be a necessity for keeping calves healthy, and thus for maintaining the profitability of the enterprise.

Under particular environmental and nutritional conditions it has been conclusively shown that certain antibiotics, particularly chlortetracycline (aureomycin) and oxytetracycline (terramycin) have a beneficial effect in reducing mortality and increasing the growth rate of young calves. When colostrum is not available, owing to the death of the dam or owing to the removal of large volumes of the prepartum secretion (see Volume 1, p. 48), 250 mg chlortetracycline daily given orally for the first 5 days of life, followed by 125 mg/day for the next 16 days, will give protection from death in most cases.[1] The first dose should be given as soon after birth as possible and, for extra safeguard, 200 mg tetracycline (achromycin) should be given by intramuscular injection at the same time.

When there is a high incidence of *Escherichia coli* infections (see p. 119) in purchased calves, the prophylactic use of the tetracycline antibiotics at a level of 125 mg/day during the first 3 weeks of life may reduce the mortality rate from *E. coli* localised intestinal infection[2] even with calves given adequate amounts of colostrum.

E. coli is resistant to the action of natural penicillin. For instance, procaine penicillin at a level of 250 mg/day for the first 5 days of life, followed by 125 mg/day for the next 16 days not only had no protective value against *E. coli* septicaemia, but appeared to antagonise the beneficial effect of chlortetracycline if fed in combination with it.[1]

On the other hand, the synthetic penicillin, ampicillin, has a range of antibacterial activity similar to that of the tetracyclines. In particular, it has been used as a prophylaxis against salmonellosis, as it is rapidly absorbed after oral administration and high concentrations are found in the bile and urine. At a level of 50–200 mg/day for a fortnight,[3] ampicillin prevented the development of diarrhoea and increased the weight gains of calves exposed to *Salmonella dublin* infection. It has been suggested that the particular beneficial effect of ampicillin is associated with its high concentration in the bile, and the fact that, in chronic infections, salmonella organisms are likely to be localised in the gall bladder.[4]

The inclusion of tetracycline antibiotics in concentrate mixtures at a low level, so that the calf consumes about 15–80 mg antibiotic daily, has often been found to increase food intake and efficiency of food conversion.[5] Indeed, the routine feeding of antibiotics in concentrate mixtures to calves weaned on to dry food at 3 weeks of age has been considered essential for the success of the system.[6] The improvement in voluntary intake of food and in weight gain, resulting from their inclusion, has been attributed[7] to a reduction in the rate of fermentation and of gas production in the rumen, with the resultant possibility that a greater proportion of the diet is digested in the abomasum and intestine. Higher blood glucose levels in ruminant calves given chlortetracycline give some support to this theory.[8–13] It has also been suggested that the beneficial effects of certain antibiotics in the young calf occur because they make pathogenic bacteria more sensitive to phagocytosis.[14–16]

Increased resistance to antibiotics, of strains of *E. coli* isolated from calves that have received antibiotics, is well known.[17–19] Moreover, these strains may pass to other calves in the community that have received no antibiotics. A survey in the U.S.A.,[20] where tetracyclines have been used in calf foods for a number of years, showed that, of fifty-six cultures from nine 'O' groups isolated from calves, forty-seven and fifty-one cultures were insensitive (minimum inhibitory concentration $> 10\ \mu g/ml$) to chlortetracycline and oxytetracycline, respectively, whereas only two were insensitive to furazolidone. In 1959, the phenomenon of infectious resistance was reported from Japan,[21,22] and confirmed in the U.K.[23] This form of resistance differs from mutational resistance in that the resistance can be transmitted from one organism of a species to another organism of the same or different species by contact. A high incidence of infectious resistance has been found in strains of *E. coli* isolated from calves. In one experiment,[24] 65% of strains of *E. coli*, isolated from calves that had died from or were suffering from diarrhoea, were resistant to more than one drug, and twenty-two out of twenty-four strains showed infective resistance. Similarly, in another experiment,[25] twenty out of forty-five multiple resistant strains (resistant to at least three drugs, including tetracycline) isolated from calves, were able to transfer their resistance to *Salmonella typhi-murium*, although only four out of forty-five strains transferred their resistance to *E. coli*. With calves having low viable *E. coli* counts in their faeces, it appears to be more difficult to find strains of *E. coli* having transferable drug resistance, than with calves excreting faeces containing a high *E. coli* count.

However, there is little unequivocal evidence[26] that routine antibiotic feeding in conventional rearing of dairy herd replacements has resulted in an increased incidence of diseases that have become more difficult to control. More evidence,[27] although mainly subjective, indicates that under very intensive systems of liquid feeding for veal production, routine antibiotic feeding may result in a decline in the effectiveness of the antibiotic.

Under the Therapeutic Substances (Supply of Antibiotics for Agricultural Purposes) Regulations 1953, antibiotic supplements may not be included in proprietary calf foods, but a farmer may obtain an antibiotic supplement on a veterinary prescription. The Report of the Joint Committee of the A.R.C. and M.R.C. on Antibiotics in Animal Feeding[28] recommended that 'farmers be allowed to purchase supplemented calf feed, suitably labelled, from compounders and that future regulations to this end should limit supplementation to an officially determined level of antibiotic and should make clear that the feed was to be given to only young calves'. Since that report was published, the new evidence on multiple and infectious resistance has resulted in a deferment of new regulations to allow the use of antibiotics in calf foods.

In general, it would seem wiser for a farmer to modify his husbandry practice to reduce the risk of infection rather than to resort to the routine use of antibiotics. The composition of some of the crude antibiotic supplements that are available is as follows:

Terramycin TM5:	11 g oxytetracycline/kg (5 g/lb)
Aurofac 2A:	8 g chlortetracycline/kg (3·6 g/lb)
Aurofac D (for inclusion in liquid diets):	11 g chlortetracycline/kg (5 g/lb)

REFERENCES

1. INGRAM, P. L., SHILLAM, K. W. G., HAWKINS, G. M. and ROY, J. H. B. *Br. J. Nutr.* **12**, 203 (1958)
2. LANG, J. M., ROY, J. H. B., SHILLAM, K. W. G. and INGRAM, P. L. *Br. J. Nutr.* **13**, 463 (1959)

3. KERR, W. J. and BRANDER, G. C. *Vet. Rec.* **76**, 1105 (1964)
4. ACRED, P., LARKIN, P. J. and MIZEN, L. *Vet. Rec.* **79**, 103 (1966)
5. LASSITER, C. A. *J. Dairy Sci.* **38**, 1102 (1955)
6. PRESTON, T. R., MACLEOD, N. A. and DINDA, P. K. *Anim. Prod.* **1**, 13 (1959)
7. PRESTON, T. R. *Wld Rev. Nutr. Diet.* **4**, 117 (1963)
8. MURLEY, W. R., ALLEN, R. S. and JACOBSON, N. L. *J. Anim. Sci.* **10**, 1057 (1951)
9. HIBBS, J. W., CONRAD, H. R. and POUNDEN, W. D. *J. Anim. Sci.* **11**, 764 (1952)
10. HIBBS, J. W., CONRAD, H. R. and POUNDEN, W. D. *J. Dairy Sci.* **37**, 724 (1954).
11. VANDERSALL, J. H., HIBBS, J. W. and CONRAD, H. R. *J. Dairy Sci.* **39**, 929 (1956)
12. EVERETT, J. P. Jr., BROWN, L. D., LASSITER, C. A., JACOBSON, D. R. and RUST, J. W. *J. Dairy Sci.* **41**, 1407 (1958)
13. DINDA, P. K. 'Some effects of chlortetracycline on the nutrition of the early weaned calf' *Ph.D. Thesis* (University of Aberdeen, 1960)
14. RADISSON, J. J., SMITH, C. K. and WARD, G. M. *J. Dairy Sci.* **39**, 1260 (1956)
15. RADISSON, J. J., BARTLEY, E. E., LORD, T. H. and SWENSON, M. J. *J. Dairy Sci.* **39**, 1386 (1956)
16. MACFADDEN, D. L. and BARTLEY, E. E. *J. Anim. Sci.* **18**, 1378 (1959)
17. SMITH, H. W. and CRABB, W. E. *Vet. Rec.* **68**, 274 (1956)
18. SMITH, H. W. and CRABB, W. E. *J. gen. Microbiol.* **15**, 556 (1956)
19. INGRAM, P. L. and LOVELL, R. *Vet. Rec.* **72**, 1183 (1960)
20. GLANTZ, P. J. *Cornell Vet.* **52**, 552 (1962)
21. OCHIAI, K., YAMANAKA, T., KIMURA, K. and SAWADA, O. *Nippon Iji Shimpo* no. 1861, **34** (1959)
22. AKIBA, T., KOYAMA, K., ISHIKI, Y., KIMURA, S. and FUKUSHIMA, T. *Jap. J. Microbiol.* **4**, 219 (1960)
23. ANDERSON, E. S. *Br. med. J.* **ii**, 1289 (1965)
24. SMITH, H. W. *Vet. Rec.* **80**, 464 (1967)
25. WALTON, J. R. *Lancet* **ii**, 1300 (1966)
26. GOULD, C. M. *Vet. Rec.* **70**, 697 (1958)
27. GRAY, W. V. *Vet. Rec.* **74**, 628 (1962)
28. Agricultural Research Council and Medical Research Council. *Report of the Joint Committee on Antibiotics in Animal Feeding* (H.M.S.O. London, 1962)

CHAPTER SEVEN

Milk Substitute Diets

'HIGH FAT' VERSUS 'LOW FAT' DIETS

Before 1960, the majority of milk substitutes contained about 1% fat in the dry matter, and consisted basically of milk by-products and plant products. The calorific value of these reconstituted diets was only about half that of whole milk. As a result of the increase in intensity of production, the evidence that starch and its degradation products were poorly digested (see p. 6), and a world surplus of fats, milk substitutes containing 15–20% fat began to be produced commercially. Experiments on the inclusion of such fats had been made for many years, particularly in the U.S.A.

The evidence suggests that the feeding of high-fat milk substitutes results in increased fat deposition in the carcase, but the dietary fat does not appear to be a readily available source of energy for increasing muscle and bone deposition under optimum environmental conditions. Thus, at *ad lib.* levels of intake, the use of a high-fat milk substitute appears to result in the same protein and bone growth, but greater fat deposition than feeding a diet of skim milk. If fat deposition at an early age is definitely required, as in veal production, the advantage of feeding a high-fat diet is clear cut. However, there appear to be three other advantages that accrue from the use of high-fat milk substitutes for all forms of calf rearing. First, there is a lower risk of diarrhoea occurring in calves given these diets, either because the fat itself is costive or because it serves to reduce the level of the protein and soluble carbohydrate in the diet, an excess of which may have detrimental effects (see p. 123). Secondly, if a high-fat milk substitute is fed at a low level to calves, with the intention of encouraging the consumption of dry food, reserves of fat are probably laid down that may be drawn upon after weaning at a young age; at this time the calf is receiving a reduced energy supply due to the change from monogastric to ruminant digestion. Most of the evidence for the advantage of high-fat milk substitute diets in early weaning systems indicates an increased performance after weaning.[1] Thirdly, high-fat milk substitutes are essential if automatic feeding systems are used.

Milk substitutes consist basically of skim milk powder and vegetable or animal fat, although a proportion of buttermilk powder and whey powder is often included. A small proportion of glucose, soya bean flour and cereal flour (1–3%) may also be added, together with certain minerals and vitamins. If a good quality milk substitute is used, there should be no necessity for the feeding of any whole milk after the colostrum-feeding period.

QUALITY OF SKIM MILK POWDER

The quality of milk substitute diets given directly after the colostrum-feeding period is an important factor in predisposing calves to an *E. coli* localised intestinal infection (see p. 120). As milk substitutes are based to a large extent on skim milk powder, which usually comprises at least 50% of the total dry matter, the quality of the skim milk powder is of considerable importance. If in the processing of skim milk, denaturation of the whey proteins has occurred to a marked extent, then such milk powders will have a detrimental effect on the calf. Associated with the denaturation of the whey proteins is a reduction in ionisable calcium,[2–10] the release of SH groups,[11–16] poor clotting ability by rennet and reduced digestibility, but no effect on the biological value of the protein.[17] The effect of such diets may be shown in three ways.

1. In the absence of an active infective agent, weight gain may be reduced by as much as 30% in the first 3 weeks of life.[17]
2. The rate of build up of infection (see p. 123)[18] will be greater when large numbers of susceptible calves are passed through a calf house.
3. The incidence of diarrhoea and mortality will be higher once the 'infection' has been built up,[19] as evidenced by the dominance of one or two strains of *E. coli*.[20]

Some of the common processing methods, together with their effect on the whey proteins are given in Table 7.1. Only a mild preheating treatment before spray-drying gives a product closely similar to that of fresh milk. The content of non-casein nitrogen (a measure of the whey proteins, together with a small amount of proteose-peptone nitrogen and non-protein nitrogen which are little affected by heat), as a percentage of the total nitrogen content indicates how much denaturation of the whey protein has occurred during processing. These analyses are meaningful if done on skim milk or skim milk with added fat but not if other products, such as dried whey or vegetable protein are added. In raw milk the percentage is about 25, but in U.H.T. sterilised milk it is only about 11. The

Table 7.1 EFFECT OF HEAT TREATMENT OF MILK ON THE CALF[21, 22]

Treatment of milk	*Per cent total milk nitrogen as non-casein nitrogen and effect on calf*
	No demonstrable effect on calf
Raw	25
Holder pasteurised (63°C for 30 min)	23
Spray-dried skim (preheating temperature 77°C for 15 sec)	22
	Detrimental to the calf especially during the first 3 weeks of life
Spray-dried skim (preheating temperature 74°C for 30 min +)	15
Roller-dried skim (110°C)	13
U.H.T. sterilised (135°C for 1–3 sec)	11

inclusion of whey in the milk substitute will invalidate the results, because of its high concentration of proteose-peptone N and non-protein N; in dried whey, the majority of the proteins are usually denatured.

Antibiotic supplementation does not appear to prevent the detrimental effect of a severely heat-treated skim milk.[17] Although it is known that the reduction in calcium content of a 'synthetic milk' from 1·26 to 0·73 g/l. will induce diarrhoea,[23] only a slight beneficial effect of calcium addition to a severely heat-treated milk has been found, even though such addition improved the clotting of the milk *in vitro.*[24]

As might be expected from the finding that there is no change in biological value, supplementation with the amino-acid methionine is also ineffective. Addition of undenatured whey proteins from liquid whey appears to be the only factor that will improve a severely-heated skim milk powder,[25] this finding is in keeping with other results which showed that the presence of denatured whey proteins increases the time taken for the clotting of the casein of milk by rennet.[26] The presence of denatured whey proteins thus appears to have an important detrimental effect on the rate of clotting, and this is additional to the fact that heat treatment sufficient to cause denaturation also results in the binding of the calcium and a reduction in the concentration of ionisable calcium.

Diarrhoea in calves has been associated with the rapid passage of milk from the abomasum without sufficient gastric digestion.[27] It is also known that, after feeding, there is a much more rapid fall in pH in the abomasum and a much lower proportion of undigested protein passing into the duodenum of a calf given a diet containing a 'mildly' preheated spray-dried skim milk, than when a diet containing a 'severely' preheated spray-dried skim milk is given.[28]

It must be emphasised that the most marked detrimental effect of a poor quality skim milk powder occurs during the first 3 weeks of life under conditions of sub clinical infection with *E. coli*, and that the breed of calf may also be important; calves of the Friesian breed may well be less susceptible than those of smaller breeds, possibly due to their higher potential for digesting protein.

It has been suggested in Holland[29,30] that, due to poor rennet clotting of milk substitute diets, absorption in the small intestine of only partially broken down milk proteins may occur, with the resultant formation of antibodies, so that at a later stage anaphylactic shock symptoms may be observed. This suggestion has not been confirmed by others, who consider that some factor in the milk substitute, other than the milk protein, was involved.[31] More recently, it has been reported that calves given a synthetic milk containing 10% soya bean meal produced anti-soya bean protein antibodies. No allergic reactions were observed but growth rate was poor.[32]

WHEY POWDER, LACTOSE AND GLUCOSE

The inclusion of dried whey in milk substitutes is usually limited to less than 20% of the dry matter, owing to the high incidence of diarrhoea resulting from larger amounts. The inclusion of whey results in a reduction in the protein content of the milk substitute, which may become too low in protein for maximum growth rate during the first 8 weeks of life. The increase in diarrhoea, as a result of feeding calves with large amounts of whey, has been associated in part with the high mineral content of the whey, which appears to accelerate the rate of passage from the abomasum.[33,34] Additional fat has been shown to have an ameliorating effect; this effect is considered to be due to the fat delaying the passage of ingesta from the abomasum. However, other workers have found either an increased passage from the abomasum with additional fat or no consistent effect on the rate of flow from the abomasum.[35,36]

When more than 500 g of lactose or glucose are given daily, there is a tendency for the calf to have diarrhoea[27] (see p. 127). This appears to be especially so when the amount of fat in the diet is low, for greater amounts of lactose can be ingested by a calf without ill effect when contained in whole milk given *ad lib.* In France,[37] a comparison of liquid diets containing 2·3–6·9% glucose and 0·5–3·5% fat, showed that the highest weight gains were obtained when the highest concentration of glucose and of fat was used. Although retention of nitrogen was highest at the highest level of glucose, diarrhoea was more frequent; this finding led to the recommendation that milk substitutes should contain not more than 2·3% glucose and a fat content of 2·5–3·5%.

FAT

The requirement of fat is probably quite low, for calves can be reared solely on a skim milk diet with additional vitamins A and D after the colostrum-feeding period, to achieve quite reasonable growth rates in the first 3 months of life.[38] However, growth rates will be lower than with a diet of whole milk and there will be a tendency for more diarrhoea, since butterfat has been shown to have a costive effect.

The addition of vegetable or animal fat to liquid skim milk or reconstituted dried skim milk is normally beyond the resources of the farmer, for, unless correctly emulsified so that the fat particle size is less than 3–4 μ in diameter [(3–4/1 000) mm], such diets tend to cause diarrhoea and loss of hair during the first 3 weeks of life.[23,39,40]

METHOD OF INCORPORATION OF FAT

Fat may be incorporated either by homogenisation into liquid skim milk, followed if necessary by drying, or by blending into skim milk powder with the aid of an emulsifying agent. In general, homogenisation of the fat into liquid skim milk, prior to drying, will give the best quality product.[41] With homogenisation, the mean fat globule size should not be more than 3–4 μ, whereas the blended globule is usually 10–20 μ in diameter. From the results of blood lipid levels, it would appear that older calves can probably tolerate fat globules of a size greater than 3–4 μ without ill effect.[41] Indeed,one study of the performance of calves given milk substitute diets containing fat globules of 2·5 μ or 20 μ in size produced only a marginal improvement in performance when the diet containing the smaller sized fat globules was given.[42] Moreover, with

2–3-week old calves, emulsification of coconut fat with crude soya bean lecithin resulted in a very large increase in digestibility, which was not enhanced by homogenisation.[43] Nevertheless, for the young calf and for calves of the smaller breeds, fat globules of small size are more efficiently digested than those of larger size.

If the fat is homogenised, there should be no necessity to add an emulsifying agent, unless the fat being used is of low inherent digestibility. For instance, the addition of soya bean lecithin, as well as homogenisation, may result in increased digestibility of beef tallow,[44,45] but such addition had no effect on the digestibility of a margarine containing 39% groundnut oil, 24% palm oil and 37% coconut oil.[41] Soya bean lecithin is usually included at the rate of 1–6% of the dry matter of the diet. There are some indications that lecithin may have a beneficial effect in supplying phospholipids, as well as acting as an emulsifying agent.[46,47]

In addition to the normal emulsifying agents used in milk substitutes, such as lecithin and glyceryl monostearate, recent experiments have indicated some beneficial effects of the inclusion of saccharoglycerides, together with lecithin, on the utilisation of beef tallow during the first few weeks of life.[48–51] Saccharoglycerides are produced as the result of the transesterification of a triglyceride of natural origin with sucrose. It has been claimed in particular that saccharoglycerides improve the digestibility of palmitic and stearic acids, which have a low inherent digestibility for young calves, and also reduced the incidence of diarrhoea.[52] However, with a milk substitute containing 20% fat (4 : 1 ratio of lard to coconut oil), no difference occurred in the performance of calves given homogenised diets containing saccharoglycerides rather than lecithin.[53] From the evidence so far available, it appears that the advantages of using saccharoglycerides in homogenised milk substitutes containing tallow are marginal.

In general, where the cheaper blending method is adopted, fats of high inherent digestibility are used, whereas with homogenisation, the cheaper fats such as beef tallow are suitable.

NUTRITIVE VALUE OF FATS

There is considerable variation in the digestibility of different fats by calves, but the reasons for these differences are not clear cut. Much of the variation can be accounted for by the chain length of the constituent fatty acids rather than by their degree of saturation; thus, the shorter the chain, the greater the digestibility.[54] However, other workers have found little evidence that the overall fatty acid composition had any marked effect on digestibility and suggested that the digestibility of a particular fat was related to its triglyceride make-up (e.g. the proportion

Table 7.2 DIGESTIBILITY (%) OF VARIOUS FATS, INCLUDED IN MILK SUBSTITUTES AT A LEVEL OF 20% OF THE DRY MATTER, IN COMPARISON WITH THAT OF BUTTERFAT

Age	*4 weeks*	*10 weeks*
Butterfat	97	97
Margarine	97	96
Refined lard	93	93
Equal parts refined lard and beef dripping	89	92
Beef dripping	85	89

of glyceryl tristearate in the total triglycerides) rather than the relative amounts of each fatty acid present.[55]

Another viewpoint is that fats with melting points below body temperature are better utilised than those with higher melting points; tristearin with a melting point of 71·6–73·2°C is considered to be virtually not assimilated.[56] On the other hand, it has been suggested that, although tallow is of much lower digestibility than most vegetable fats, the utilisation and thus deposition of fat in the carcase is considerably greater with a diet containing tallow.[52]

Digestibility of beef tallow tends to be lower than that of lard,[22, 57] which in turn is lower than that of margarine. Some values for digestibility of fats are given in Table 7.2. With fats of low digestibility, there is an increase in digestibility with age;[22, 55, 57-59] this appears to be associated with a reduction in the faecal excretion of neutıal fat and an increase in the amount of saponified fat.[42, 60] In this connection, the use of a fat of low digestibility may increase the requirement of calcium and magnesium, owing to the formation of soaps. In a comparison of milk substitutes containing beef tallow and fat rendered from beef bones (marrow fat), an increased amount of soaps was produced in the faeces on the former diet and this finding was associated with reduced serum magnesium levels.[42]

The original work with vegetable oils showed that homogenised milk substitute diets containing 3·5% cottonseed, soya bean or maize oil gave very poor results, whereas lard and tallow gave moderately good results and coconut oil gave an intermediate response.[61] However, these results may have been distorted by the existence of a subclinical vitamin E deficiency. The calves given the cottonseed, soya bean or maize oil diets showed a loss of hair, first from the face, ears and neck and later from the cannons of the rear legs.

Alopecia in calves has also been associated with unhomogenised diets.[39, 41] It is generally accepted that this loss of hair occurs as a result of direct contact of the hair with the diet or faeces. On the other hand, it has been suggested that hair loss and development of eczema on unhydrogenated soya bean oil diets is due to the secretion of oil through the skin, hydrogenated soya bean oil having much less effect.[62] Similarly, loss of hair has been reported when milk substitutes containing 20% whale oil, or a mixture of 9·3% hardened palm and 6·2% soya oil have been given, the condition being attributed to the absorption of unphysiological fat components which are excreted through the sebaceous glands.[63]

Even when homogenised, unhydrogenated soya bean oil has been found to cause a poor rate of growth, rough hair coats, and a high incidence of diarrhoea and mortality, but diets containing hydrogenated soya bean oil have given results nearly as good as diets containing butter oil.[64–69] No difference has been found in the digestibility of palm and palm kernel oil, whether hydrogenated or not, digestibility being about 90% compared with 97% for butterfat.[70–72]

Although lard and tallow are comparatively well utilised by the calf,[65, 73–83] there is a tendency for a higher incidence of diarrhoea with these fats[22, 84] than with butterfat during the first 14 days of life. Coconut oil is often included, together with tallow, to supply short-chain fatty acids, and there are indications that digestibility of the tallow is thereby increased. Coconut oil alone has been shown to cause severe diarrhoea.[85, 86] In Europe, hydrogenated herring and fish oil have been found satisfactory,[85, 87] as has sunflower oil which is considered to be superior to fish oil.[88]

It is essential that an antioxidant should be included in milk substitutes containing additional fat, if rancidity is to be prevented. This is especially so with fats containing a high proportion of unsaturated fatty acids. The requirement of the essential fatty acids (linoleic, linolenic and arachidonic) for calves is not known, although it has been claimed that deficiency symptoms, including dandruff, dull haircoat and partial alopecia, have been produced in calves over 6 weeks of age on fat-free diets.[47]

QUANTITY OF FAT

The quantity of fat that should be included in a milk substitute diet depends on the digestibility of the fat that has been chosen, the amount of body fat deposition that is required, and the relative costs of fat and skim milk. A liquid diet containing 4% tallow will be unsatisfactory for the young calf, whereas one containing 4% margarine with a digestibility similar to that of butterfat can be used directly after the colostrum-feeding period. On the other hand, if the content of tallow in the diet is 2–3%, it may be used directly after the colostrum-feeding period, provided that the tallow is of good quality; beef fat can be obtained in various grades from tallow to beef dripping.

As mentioned earlier, the inclusion of high levels of fat will have little effect on protein deposition, but will markedly increase fat deposition. For instance, a milk substitute diet containing 4·2% margarine fat had little effect on protein deposition compared with one containing 2·8% fat, but resulted in a 50% increase in peri-renal fat deposition.[89] In Australia, the most effective feed conversion was obtained with skim milk containing 2% rather than 3% tallow.[90]

EFFECT OF DIETARY FAT ON QUALITY OF CARCASE FAT

The quality of the dietary fat no doubt influences the composition of the carcase fat, but only to a limited degree. For instance, linseed oil included in a milk substitute increased the proportion of linoleic, linolenic and arachidic acids, but reduced the proportion of oleic acid.[91] Similarly, a higher proportion of these acids was deposited in the fat of calves given a milk substitute containing 14% coconut oil and 4% soya oil, than in the fat of those given a milk substitute containing 18% tallow.[52] In spite of the higher content of lauric acid in the vegetable oil, negligible amounts were deposited in the fat, the rest being metabolised or excreted through the skin.

COMPOSITION OF MILK SUBSTITUTE DIETS

At the present state of knowledge, a milk substitute of the following composition should supply the dietary needs of the calf when used as a sole diet:

Animal or vegetable fat:	17–20%
Crude soya bean lecithin:	1–2%
Skim milk solids:	78–82%

For the best quality product, the fat should be homogenised into liquid skim milk, and then spray-dried, using a mild preheating treatment. 10–15% of dried whey powder could be included in place of an equivalent amount of skim milk solids, if maximum live weight gain from birth is not required.

The following additional minerals and vitamins (see Chapters 3–5) are recommended:

Magnesium:	250 mg/kg dry matter (113 mg/lb)
Iron:	75–100 mg/kg dry matter (34–45 mg/lb)
Manganese:	40 mg/kg dry matter (18 mg/lb)
Copper:	10 mg/kg dry matter (4·5 mg/lb)
Cobalt:	100 μg/kg dry matter (45 μg/lb)
Zinc:	12 mg/kg dry matter (5·4 mg/lb)
Iodine:	120 μg/kg dry matter (54 μg/lb)
Vitamin A:	12 000–20 000 i.u. (3·6–6·0 mg)/kg dry matter [5 000–9 000 i.u. (1·6–2·7 mg)/lb]
Vitamin D:	1 800–3 200 i.u./kg dry matter (815–1 450 i.u./lb)
Vitamin E:	20 mg/kg dry matter (9 mg/lb)
Vitamin B_{12}:	30 μg/kg dry matter (14 μg/lb)

REFERENCES

1. GRIFFITHS, T. W. and McGANN, P. S. *Anim. Prod.* **8**, 349 (1966)
2. LAMPITT, L. H. and BUSHILL, J. H. *Biochem. J.* **28**, 1305 (1934)
3. VERMA, I. S. and SOMMER, H. H. *J. Dairy Sci.* **33**, 397 (1950)
4. BERNADONI, E. A. and TUCKEY, S. L. *J. Dairy Sci.* **33**, 409 (1950)
5. HARMAN, T. D. and SLATTER, W. L. *J. Dairy Sci.* **33**, 409 (1950)
6. HILGEMAN, M. and JENNESS, R. *J. Dairy Sci.* **34**, 483 (1951)
7. BAKER, J. M., GEHRKE, C. W. and AFFSPRUNG, H. E. *J. Dairy Sci.* **37**, 643 (1954)
8. CHRISTIANSON, G., JENNESS, R. and COULTER, S. T. *Analyt. Chem.* **26**, 1923 (1954)
9. HOSTETTLER, H. and STEIN, J. *Landw. Jb. Schweiz* **7**, 163 (1958)
10. DAVIS, D. T. and WHITE, J. C. D. *Int. Dairy Congr. XV, London* **3**, 1677 (1959)
11. GOULD, I. A. Jr. and SOMMER, H. H. *Tech. Bull. Mich. agric. Exp. Stn* no. 164 (1939)
12. JOSEPHSON, D. V. and DOAN, F. J. *Milk Dlr* **29**, 35 (1939)
13. TOWNLEY, R. C. and GOULD, I. A. *J. Dairy Sci.* **26**, 843 (1943)
14. LARSON, B. L. and JENNESS, R. *J. Dairy Sci.* **33**, 896 (1950)
15. HUTTON, J. T. and PATTON, S. *J. Dairy Sci.* **35**, 699 (1952)
16. ZWEIG, G. and BLOCK, R. J. *J. Dairy Sci.* **36**, 427 (1953)
17. SHILLAM, K. W. G. and ROY, J. H. B. *Br. J. Nutr.* **17**, 171 (1963)
18. ROY, J. H. B., PALMER, J., SHILLAM, K. W. G., INGRAM, P. L. and WOOD, P. C. *Br. J. Nutr.* **9**, 11 (1955)
19. SHILLAM, K. W. G., ROY, J. H. B. and INGRAM, P. L. *Br. J. Nutr.* **16**, 585 (1962)
20. WOOD, P. C. *J. Path. Bact.* **70**, 179 (1955)
21. SHILLAM, K. W. G. and ROY, J. H. B. *Int. Congr. Anim. Prod. VIII, Hamburg* **3**, 276 (1961)
22. ROY, J. H. B. *Vet. Rec.* **76**, 511 (1964)
23. KASTELIC, J., BENTLEY, O. G. and PHILLIPS, P. H. *J. Dairy Sci.* **33**, 725 (1950)

24. SHILLAM, K. W. G. and ROY, J. H. B. *Br. J. Nutr.* **17**, 183 (1963)
25. SHILLAM, K. W. G., ROY, J. H. B. and INGRAM, P. L. *Br. J. Nutr.* **16**, 593 (1962)
26. KANNAN, A. and JENNESS, R. *J. Dairy Sci.* **44**, 808 (1961)
27. BLAXTER, K. L. and WOOD, W. A. *Vet. Rec.* **65**, 889 (1953)
28. TAGARI, H. and ROY, J. H. B. *Br. J. Nutr.* **23** (in the press) (1969)
29. FRENS, A. M., GRIFT, J. van DER and DAMMERS, J. *Tijdschr. Diergeneesk.* **86**, 255 (1961)
30. FRENS, A. M. *Tijdschr. Diergeneesk.* **86**, 1636 (1961)
31. BOOGAERDT, J. and van KOETSVELD, E. E. *Tijdschr. Diergeneesk.* **86**, 1287 (1961)
32. van ADRICHEM, P. W. M. and FRENS, A. M. *Tijdschr. Diergeneesk.* **90**, 525 (1965)
33. OWEN, F. G., JACOBSON, N. L., ALLEN, R. S. and HOMEYER, P. G. *J. Dairy Sci.* **41**, 662 (1958)
34. BUSH, L. J., SCHUH, J. D., TENNILE, N. B. and WALLER, G. R. *J. Dairy Sci.* **46**, 703 (1963)
35. ESPE, D. L. and CANNON, C. Y. *J. Dairy Sci.* **18**, 141 (1935)
36. ASH, R. W. *J. Physiol. Lond.* **172**, 425 (1964)
37. MATHIEU, C. M. and TUGNY, H. de *Annls Biol. anim. Biochim. Biophys.* **5**, 21 (1965)
38. CONVERSE, H. T. *Circ. U.S. Dep. Agric. no.* 822 (1949)
39. BATE, W., ESPE, D. and CANNON, C. Y. *J. Dairy Sci.* **29**, 41 (1946)
40. HUFF, J. S., WAUGH, R. K. and WISE, G. H. *J. Dairy Sci.* **34**, 1056 (1951).
41. ROY, J. H. B., SHILLAM, K. W. G., THOMPSON, S. Y. and DAWSON, D. A. *Br. J. Nutr.* **15**, 541 (1961)
42. THOMKE, S. *Züchtungskunde* **35**, 265 (1963)
43. HOPKINS, D. T., WARNER, R. G. and LOOSLI, J. K. *J. Dairy Sci.* **42**, 1815 (1959)
44. RAVEN, A. M. and ROBINSON, K. L. *J. Sci. Fd Agric.* **4**, 214 (1964)
45. ROY, J. H. B., STOBO, I. J. F. and GASTON, H. J. *Rep. natn. Inst. Res. Dairy* p. 43 (1965)
46. HAMADA, T., KAMEOKA, K. and MORIMOTO, H. *Agric. biol. Chem.* **29**, 998 (1965)
47. LAMBERT, M. R., JACOBSON, N. L., ALLEN, R. S. and ZALATEL, J. H. *J. Nutr.* **52**, 259 (1954)
48. JOUSSELIN, W., LADRAT, J., CRAPLET, C. and BARRÉ, P. *Bull. Acad. vét. Fr.* **37**, 277 (1964)
49. Le PROVOST, F. *Les Saccharoglycérides. Thesis D Vét Med.* (Ecole Nationale Véterinaire de Toulouse, 1965)
50. LADRAT, J. and JOUSSELIN, W. *Bull. Acad. vét. Fr.* 38, 497 (1965)
51. LADRAT, J. and JOUSSELIN, W. *Bull. Acad. vét. Fr.* 38, 501 (1965)
52. AMICH GALI, J. 'Research on the value of animal fats and sugar glycerides in the diets of calves.' (National Renderers Association, Inc., Des Plaines, Illinois, 1965)
53. CATALANO, A. *Alimentaz anim.* **10**, 147 (1966)
54. LLOYD, L. E. and CRAMPTON, E. W. *J. Anim. Sci.* **16**, 377 (1957)
55. RAVEN, A. M. and ROBINSON, K. L. *J. Soc. Dairy Technol.* **17**, 5 (1964)
56. KONDYREV, V. E. *Whole milk replacers in calf rearing.* ('Kolos' Publishing House, Moscow, 1964).
57. BRÜGGEMANN, J. and BARTH, K. *Z. Tierphysiol. Tierernähr.* **14**, 284 (1959)
58. CUNNINGHAM, H. M. and LOOSLI, J. K. *J. Dairy Sci.* **37**, 453 (1954)
59. LASSITER, C. A., DUNCAN, C. W. and CHRISTIE, L. D. *Q. Bull. Mich. St. Univ. agric. Exp. Stn* **40**, 282 (1957)

60. THOMKE, S. *Züchtungskunde* **35**, 214 (1963)
61. GULLICKSON, T. W., FOUNTAINE, F. C. and FITCH, J. B. *J. Dairy Sci.* **25**, 117 (1942)
62. de MAN, T. J. *Tijdschr. Diergeneesk*, **76**, 175 (1951)
63. GRÜNDER, H. F. and MUSCHE, R. *Dt. tierärztl. Wschr.* **69**, 437 (1962)
64. JACOBSON, N. L. and CANNON, C. Y. *J. Dairy Sci.* **30**, 587 (1947)
65. WIESE, A. C., JOHNSON, B. C., MITCHELL, H. H. and NEVENS, W. B. *J. Dairy Sci.* **30**, 87 (1947)
66. JACOBSON, N. L., CANNON, C. Y. and THOMAS, B. H. *J. Dairy Sci.* **32**, 429 (1949)
67. MURLEY, W. R., JACOBSON, N. L., WISE, G. H. and ALLEN, R. S. *J. Dairy Sci.* **32**, 609 (1949)
68. BARKER, H. B., WISE, G. H. and JACOBSON, N. L. *J. Dairy Sci.* **35**, 507 (1952)
69. ADAMS, R. S., GULLICKSON, T. W., GANDER, J. E. and SAUTTER, J. H. *J. Dairy Sci.* **42**, 1552 (1959)
70. RAVEN, A. M. and ROBINSON, K. L. *Br. J. Nutr.* **12**, 469 (1958)
71. RAVEN, A. M. and ROBINSON, K. L. *Br. J. Nutr.* **13**, 178 (1959)
72. RAVEN, A. M. and ROBINSON, K. L. *Br. J. Nutr.* **14**, 135 (1960)
73. BLAXTER. K. L. and WOOD, W. A. *Br. J. Nutr.* **5**, 11 (1951)
74. JOHNSON, B. C., HOPPER, J. H. and GARDNER, K. E. *J. Dairy Sci.* **36**, 599 (1953)
75. HOPPER, J. H., GARDNER, K. L. and JOHNSON, B. C. *J. Dairy Sci.* **37**, 431 (1954)
76. RITCHEY, S. J., HOPPER, J., GARDNER, K. E. and JOHNSON, B. C. *J. Dairy Sci.* **39**, 1070 (1956)
77. THOMAS, J. W. and OKAMOTO, M. *J. Dairy Sci.* **39**, 928 (1956)
78. BELL, J. M. *Can. J. Anim. Sci.* **38**, 103 (1958)
79. LARSEN, J. B. *Beretn. Forsøgslab.* no. 303 (1958)
80. COMBERG, G. and GÖLLNITZ, L. *Tierzucht* **13**, 150 (1959)
81. KLEISCH, J. and HORST, P. *Züchtungskunde* **31**, 68 (1959)
82. KESLER, E. M., WILSON, J. M. and MOORE, H. L. *Prog. Rep. Pa agric. Exp. Stn* no. 216 (1960)
83. ANDERSEN, P. E., LARSEN, J. B. and SORENSEN, M. *Landøkon. Forsøgslab. Efterårsmøde Årbog* p. 94 (1964)
84. JACOBSON, N. L., BROWN, L. R. and RATCLIFF, L. *Proc. Distillers Feed Conf.* **14**, 10 (1959)
85. HOMB, T. *Norg. LandbrHøgsk. Beretn. Føringsfors.* no. 99 (1960)
86. FALASCHINI, A., MONETTI, P. G., MORDENTI, A. and VIVIANI, R. *Riv. Zootec. Agric. Vet.* **3**, 3 (1965)
87. MARKOVIČ, P. and DVOŘÁČEK, M. *Věd. Pr. Ústřed. výzk. Úst. živoč. Vyroby Uhřiněvsi* **8**, 111 (1965)
88. KURILOV, P. S. and TANTSURA, A. D. *Trudӯ novocherk. zootekh.-vet. Inst.* **12**, 450 (1960)
89. STOBO, I. J. F., ROY, J. H. B. and GASTON, H. J. *Anim. Prod.* **9**, 267 (1967)
90. McINTYRE, K. H. and RAYNER, I. H. *Qd J. agric. Anim. Sci.* **23**, 299 (1966)
91. SIRÉN, M. J. *Life Sci.* **12**, 717 (1962)

PART 2

Calf Mortality and Diseases

CHAPTER EIGHT

Calf Mortality

Three surveys during 1936–37,[1] 1946–48 [2,3] and 1962–63[4,5] have shown very much the same incidence of mortality of young calves. In the first two surveys about 6% of calves born alive in England and Wales did not survive until the age of 6 months, whereas in Scotland the figure was about 12%. In the third survey throughout England, Wales and Scotland, it was estimated that, from 3·4 million live births, 140 000 home-bred and 37 000 purchased calves died before 1 year of age. This represented a mortality rate for home-bred calves of 5·0% to 6 months of age and 5·7% to 1 year of age. The percentage probability of death in home-bred calves was about equal to the reciprocal of age and percentage mortality up to age W (in weeks) was equal to $\log_e W + 1{\cdot}72$. In addition, abortions accounted for 2·1% and stillbirths for 3·3% of parturitions.[5] In spite of the fact that antibiotics have been available to the agricultural industry between the second and third surveys, little reduction in mortality appears to have occurred. It may be argued, therefore, that a mortality of 5% is the economic mortality rate, and that this level reflects the fact that the cost of reducing the mortality rate to about 1%, which would appear to be technically feasible and would be similar to that of infant mortality, is not at present economically attractive, and that a risk of 5% mortality can be tolerated by the farming community at present levels of input of labour and capital.

Of these deaths, just over 75% occurred during the first month of life in the earlier two surveys and 64% during the same period in the third survey, the greater majority being associated with the bacterium *Escherichia coli*. Of 2 046 calves, examined *post mortem* at Veterinary Investigation Centres in 1959–61, 46% of the deaths to 6 months of age were associated with *E. coli*, 24% with *Salmonella spp.*, 10% with poisoning (mainly lead) and 7% with virus infections.[6] In the 1962–63 survey, 70% of deaths were due to septicaemia (25%) and gastroenteritis (45%), and 11% due to respiratory disease.[4]

Mortality rates are highest in February, March and April, but there seems to be a steady increase from shortly after the beginning of the autumn calving season until the end of the spring calving season. Although the high incidence in the early spring months has been attributed to poor nutrition of the dam,[7] the evidence suggests that even with good nutrition of the dam there is a steady increase in the incidence of mortality during the winter months, because of the gradual build-up in the calf house of a dominance of certain strains of *E. coli* against which colostrum is ineffective (see Fig. 8.1).[8]

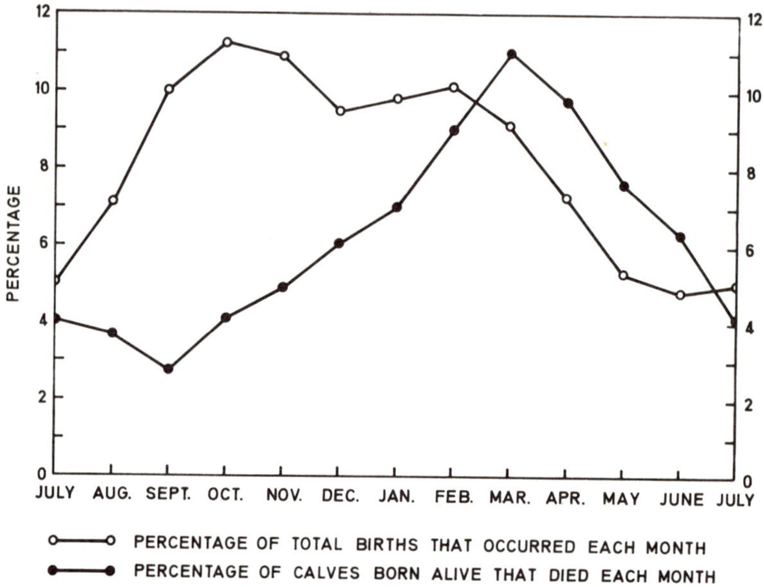

Fig. 8.1. Relationship between mortality rate and month of year. (Based on references 2 and 3)

Regional differences in the cause of mortality have been disclosed in the 1962–63 survey, with a low frequency of respiratory disorders in Devon and Cornwall and in S.E. England and S.E. Scotland, but a high frequency of alimentary disorders in S.E. Scotland, Devon and Cornwall and in Wales.[4] No regional trend in the mortality rate of home-bred calves has been found but a marked trend, from low mortality in the south to high mortality in the north of Great Britain has been shown for purchased calves.[5]

Purchased calves have a higher incidence of mortality than home-bred calves. A survey[9] of calf losses in herds using artificial insemination gave a value of 6·3% for home-bred calves, compared with 13·5% in purchased calves during the first 2 months of life in 1963–64, but the survey of 1962–63 indicated a 60% higher incidence for purchased than for home-bred calves. The main cause of death of purchased calves, however, differs from that of home-bred calves, being shifted towards salmonellosis and away from *E. coli* infection. This is not surprising, as *E. coli* infections mainly afflict calves during the first 10 days of life, during which time a large proportion of the calves will be on farms where they were born.

Bull calves have a slightly higher rate of mortality than heifers, a finding similar to that observed in human infant mortality, whereas the Channel Island breeds seem to have a higher incidence than Ayrshires, which in turn have a higher incidence than Shorthorns and Friesians. Thus the larger the average birth weight of a breed the lower seems to be the breed mortality rate. This finding may be related to differences between breeds in digestibility of milk diets (see p. 99), which may be exaggerated if the quality of the milk substitute is poor. Moreover it is much easier to overfeed a small calf than a large one. The slightly higher mortality rate within 12 hr of birth of calves sired by a Charolais bull compared with those sired by a Hereford bull, or pure bred calves[10] is probably a reflection of calving difficulties rather than differences in susceptibility to infection.

The importance of colostrum in reducing the incidence of mortality is borne out by the fact that in the survey of 1946–48 the mortality rate to 1 month of age for calves that sucked colostrum was 4%, compared with 9% for those that obtained colostrum from the bucket. This latter procedure, often adopted in Scotland, together with the fact that calves of the Ayrshire breed predominated, may in part have accounted for the higher mortality rates found in that country compared with those in England and Wales.

Similarly, in the survey of 1962–63, mortality was 69% higher for cross-bred calves and calves of the Ayrshire and Guernsey breeds given colostrum from the bucket than for those that suckled their dams. However, for the Friesian breed, the method of giving colostrum had no effect on mortality rate.

REFERENCES

1. LOVELL, R. and HILL, A. B. *J. Dairy Res.* **11**, 225 (1940)
2. WITHERS, F. W. *Br. vet. J.* **108**, 315, 472 (1952)
3. WITHERS, F. W. *Br. vet. J.* **109**, 65, 122 (1953)
4. Animal Health Division, Ministry of Agriculture, Fisheries and Food 'National Survey of Calf Wastage and Husbandry 1962–63, *Vet. Rec.* **78**, 498 (1966)
5. LEECH, F. B., MACRAE, W. D. and MENZIES, D. W. *Calf Wastage and Husbandry in Britain* 1962–63 (H.M.S.O., London, 1968)

6. Veterinary Investigation Service, Ministry of Agriculture Fisheries and Food *Vet. Rec.* **76**, 1139 (1964)
7. PAYNE, W. J. A. *Br. J. Nutr.* **3**, i (1949).
8. ROY, J. H. B., PALMER, J., SHILLAM, K. W. G., INGRAM, P. L. and WOOD, P. C. *Br. J. Nutr.* **9**, 11 (1955).
9. Milk Marketing Board, *Report of the Breeding and Production Organisation* **14**, 100 (Milk Marketing Board, Thames Ditton, Surrey 1963–64)
10. Milk Marketing Board *The Charolais Report* (Milk Marketing Board, Thames Ditton, Surrey 1966)

CHAPTER NINE

Diseases Likely to be Acquired from the Dam

BRUCELLOSIS (CONTAGIOUS ABORTION)

Calves born alive from cows suffering from *Brucella abortus* infection may themselves be infected, but such infection is only of a transient nature. As a precaution, such calves should be isolated from susceptible adults. Bull calves intended for breeding should not be fed on milk from infected cows as there is a possibility of their becoming infected in the genital organs.

To avoid an outbreak of contagious abortion in the adult herd, all heifer calves should be vaccinated with S.19 vaccine at the age of 3–6 months. Under the Ministry of Agriculture, Fisheries and Food Free Calf Vaccination Service, calves are eligible for a *free* vaccination between the 91st and 180th day of life, i.e. from 3 to 6 months of age. The use of S.19 vaccine is only permitted under the Free Calf Vaccination Service. Vaccination carried out during this period of calfhood will, in most cases, protect against abortion for at least five pregnancies. Details of the free service may be obtained from Divisional Veterinary Officers of the Ministry, and vaccination will normally be carried out by a local veterinary surgeon. The Ministry requires farmers to undertake not to vaccinate any animals outside the prescribed age without its consent. Vaccinated animals will be identified by a tattoo containing a symbol and two figures. One figure will denote the quarter and the other the year of the decade. The site of the mark will be the caudal fold under the tail.

With the intention of eventually eradicating contagious abortion, the Brucellosis (Accredited Herds) Scheme was introduced in 1967. This scheme sets up a voluntary register of brucellosis accredited herds. To enter this register, a herd is first inspected by the Ministry's staff to ensure that it would be practicable to maintain the herd in a brucella-free state if the required tests were passed.

For dairy herds, three clear Milk Ring Tests at not less than three-monthly intervals will be needed before a herd can become 'supervised'. A negative serum agglutination test of all female cattle and bulls over certain ages will then qualify the herd for full registration. Routine tests will be made at prescribed intervals thereafter.

For beef herds, a screening serum agglutination test that proves negative will allow the herd to become 'supervised' and a further clear agglutination test after a prescribed interval will qualify the herd for full registration.

Once a herd has become 'supervised', compensation will be paid for reactors at a rate of 100% subject to a maximum of £160, with the Ministry retaining the salvage value. 'Supervised' herds will be required to comply with certain conditions; the purchase of cattle from other herds will only be allowed from those of the same status and a permit will be required; adequate fencing must be provided to prevent contact with animals from herds not of the same status; and all premature calvings and abortions must be notified to the Ministry's Animal Health Division.

TUBERCULOSIS

Calves may become infected with tuberculosis before birth, although this probably occurs quite rarely. In post-natal calfhood tuberculosis, there is little doubt that the infection originates from the feeding of milk containing tubercle bacilli, and from the contact of calves with open cases of bovine tuberculosis.

Clinical symptoms of tuberculosis are usually absent and diagnosis depends essentially upon a positive mammalian tuberculin reaction. If clinical symptoms are present, they may only take the form of coughing and unthriftiness, which are common to a variety of calfhood diseases.

By careful management disease-free stock may be raised from an infected herd. Each calf should be taken away from its dam at birth; on no account should the dam be allowed to lick or suckle the calf. The calf should then be placed in a thoroughly disinfected calf pen and be given its dam's colostrum, preferably pasteurised, in a clean bucket.

After the colostrum feeding period, the calf should either be fed with milk from an attested cow or, if milk from an attested cow is not available, the milk should be pasteurised before feeding.

Pasteurisation can be carried out by heating the milk, and retaining it at a temperature of not less than 63°C (145°F) and not more than 66°C (150°F) for 30 min or at a temperature of not less than 72°C (162°F) for at least 15 sec. Colostrum can also be effectively pasteurised at the lower temperature, i.e. 63–66°C (145–150° F), without impairing its protective action, although about 15% of the whey proteins will be denatured.

At no time should the young stock use the same buildings, water troughs or pastures that have been contaminated by the adult stock; contaminated pastures are normally clear of infection after 6 months. Furthermore, the young stock should not use fields adjacent to those used by the adult cattle. If this is impracticable, a double fence, 1·8 m (6 ft) wide, should be erected between the two fields to prevent the adult cattle breathing on or licking the young stock.

Although Great Britain is now virtually free of tuberculosis in cattle, the methods that were adopted to achieve this are summarised for the benefit of those in countries where tuberculosis is still endemic.

The young stock were tuberculin tested at 3 months of age, and thereafter at six-monthly intervals. If any calves reacted to the test, they were disposed of as soon as possible. When the remaining heifers, reared under clean conditions, were pregnant the adult cows were tested and any reactors were slaughtered. As soon as two consecutive tests on all cattle on the premises had been carried out and no reactors found, the owner of the herd could apply for registration under The Tuberculosis (Attested Herds) Scheme, 1950. These two tests had to be done at an interval of not less than 60 days and not more than 12 months, and the application had to be made within 12 months of the date of the second test. The herd then became a Supervised Herd until an official test had taken place not earlier than 60 days from the date of the second of the two tests mentioned above. If the official test disclosed no reactor, the herd was registered in the Register of Attested Herds.

JOHNE'S DISEASE

This disease, caused by the organism *Mycobacterium johnei*, is very probably acquired during calfhood, although clinical symptoms, which include profuse diarrhoea, sometimes with a bubbly appearance of the dung, and severe emaciation, are not revealed until later in life, generally not until after the first calving. In addition to animals obviously infected, about 17% of apparently normal cattle carry the causal organism.

Infection of the calf can occur *in utero* if the dam has the disease in a very advanced stage.[1] Other calves may thus become infected by contact with the dung of an infected calf. Moreover, the organism has also been isolated from the udder of affected cows. It is not known whether the foetus of a cow with a latent infection may also become infected, but it has been suggested that all calves born of infected cows should be slaughtered.

Young animals appear gradually to build up an immunity, but the time that they take to become resistant varies very much between individuals. Poor nutrition of young animals also seems to be important, for the infection can be set up experimentally in undernourished animals more easily than in those on an adequate diet. Dairy cattle, especially those of the Channel Island breeds, seem to be particularly susceptible.

In infected herds, direct contact between calves and adult animals must be prevented. The calf should be removed at birth and be fed as described for calves born in herds infected with tuberculosis. Great care must be exercised in avoiding the contamination of colostrum with faeces and in preventing the transfer of faeces on the person or on equipment from the adult stock to the calf pens. Particular attention should also be directed to the cleanliness of the water supply; calves must not be allowed access to a water supply that may have been contaminated by older stock. As the Johne bacilli survive on pastures for up to one year, calves should not be turned out on to grass used by the adult stock during the previous 12 months.

Although a positive diagnosis of the disease may be made from the faeces of infected animals, since the organisms are found in about 30% of such animals that have diarrhoea, a negative result is not conclusive as the organism is likely to be passed out at irregular intervals. However, in 90% of cases showing a clinical infection, antibodies against the organism may be found in the blood. Detection of infected animals in the preclinical stage is much more difficult and cannot be done with certainty.

MASTITIS

It is quite common practice to suckle calves on cows affected with clinical mastitis. Calves should not, however, be allowed to suck cows suffering from acute mastitis, as illness and even deaths among the calves may ensue. On no account should a calf be allowed to suck a cow with a tuberculous udder or an udder infected with summer mastitis, even when there are no symptoms of acute mastitis and no gross changes in the appearance of the milk. In other types of mastitis the risk to the calf is probably slight, but where valuable calves are concerned the practice should be avoided.

REFERENCE

1. DOYLE, T. M. *Vet. Rec.* **70**, 238 (1958)

CHAPTER TEN

Diseases of Young Calves

ESCHERICHIA COLI INFECTIONS (COLIBACILLOSIS)

This is the most common disease of young calves and occurs mainly during the first 3 weeks of life.[1-3] The incidence of mortality is highest in the first fortnight of life; affected calves that survive this period often recover. There appear to be two distinct conditions,[4] (*a*) an *E. coli* septicaemia and (*b*) an *E. coli* localised intestinal infection, although intermediate forms are found on occasions. Moreover, in calves that appear to show some resistance to an *E. coli* septicaemia, death may be associated with pleurisy and peritonitis.[5]

E. COLI SEPTICAEMIA

In acute cases, there is a septicaemia caused by *E. coli* and the calf is usually found dead, or prostrated with death ensuing within a few hours. The acute cases usually occur very early in life, sometimes on the day after birth.

There is general agreement that *E. coli* is the responsible agent as a bacteraemia occurs before death.[5] The coli quite often appear to invade the tissues during the first 24–36 hr after birth, when the intestine is permeable to the transfer of intact antibodies into the blood stream. Certain serological types of *E. coli* in particular 'O' groups 15, 78, 86, 115 and 117 are associated with the disease in calves. These types accounted for 57% of the cases of septicaemia, with 078 K80 (B) being the predominant type and being associated with 36% of the infections. Agammaglobulinaemia (lack of γ-globulin in the blood serum) was found in 96% of calves with a septicaemia, and 11% of normal calves had low levels of serum γ-globulin in spite of having received colostrum.[6-9] In colostrum samples obtained within 16 miles of Reading, all contained

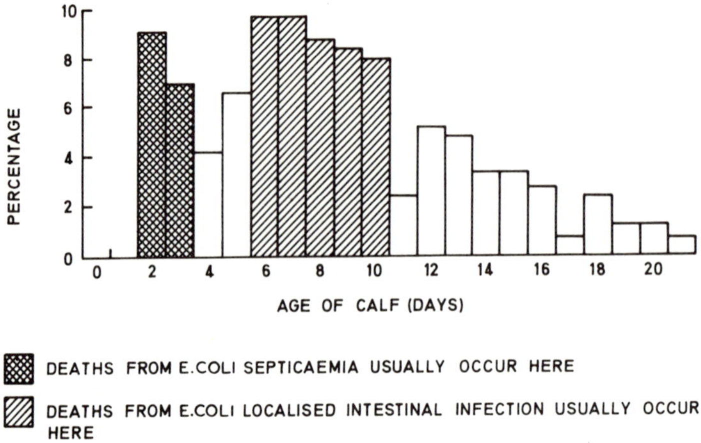

Fig. 10.1. Frequency distribution of diarrhoea during the first 3 weeks of life, showing age at which calves usually die from (a) an Escherichia coli *septicaemia and (b) an* E. coli *localised intestinal infection*

antibodies to most of the strains isolated from cases of *E. coli* septicaemia. However, no sample of colostrum contained antibodies against 078 K? (B).[10] Thus, unless a virulent strain of *E. coli* is present in the environment, the feeding of adequate colostrum during the first 24 hr of life should protect calves against an *E. coli* septicaemia.

E. COLI LOCALISED INTESTINAL INFECTION

The obvious symptom of an *E. coli* localised intestinal infection is diarrhoea and the tail becomes soiled with faeces. The colour of the faeces seems to vary in different outbreaks of the disease, in some it is yellow and watery, whereas in others it is white and pasty. In succeeding days, the appetite is variable, the diarrhoea increases and the calf becomes dehydrated, the sinking of the eyes being the most obvious symptom. If no treatment is given, the calf becomes prostrated and usually dies. This condition is usually restricted to calves that have received colostrum, and starts within the first week of life. It is associated with the presence of large numbers of *E. coli*, which form a carpet on the wall of the intestinal tract from the lower end of the small intestine up to the duodenum.[11] A similar condition may be found in older veal calves if they have been allowed to become anaemic.

At birth, there are generally no bacteria present in the contents of the alimentary tract. The first organisms to colonise the alimentary tract of the normal calf are *E. coli*, streptococci and *Clostridium welchii.* These are followed by the lactobacilli, which become the commonest organisms in the stomach and small intestine. Bacteroides are slow to colonise and are mainly restricted to the large intestine. The temporary dominance of the first organisms to colonise has been attributed to the high pH of the abomasum contents at this time, and the decline of these organisms has been associated with the subsequent drop in pH.[12]

Within 5½ hr after birth, coli are mainly restricted to the ileum and the large intestine, but by 8½ hr after birth they are found in the abomasum at a level of 10^3/ml. In the duodenum, the viable *E. coli* count reaches a peak at 1–4 days of age and in normal healthy calves it declines to a very low level at 10 days of age (10–10^2/ml), presumably because of the dominance of lactobacilli. In contrast, calves whose deaths are associated with an *E. coli* localised intestinal infection have viable counts of 10^7–10^8/ml in the duodenum at this latter age.[13]

It is not known for certain whether the presence of large numbers of bacteria actually cause the diarrhoea although this seems probable, or whether the bacteria multiply as a result of the diarrhoea. In either event, certain specific strains tend to become dominant in the intestine, although they do not usually invade the tissues further than the mesenteric lymph nodes. Diarrhoea results in the loss, in the faeces, of large amounts of water, and the electrolytes, bicarbonate, potassium, sodium

***Plate** 10.1. Calf suffering from* Escherichia coli *localised intestinal infection. Note the sunken eyes and unhealthy stance*

and chloride, in this order of magnitude, because of continued gastro-intestinal secretion and poor absorption of all nutrients, including vitamins. This results in the depletion of these electrolytes in the blood plasma and, consequently, there is movement of water and particularly of potassium from the cells of the body into the blood.[14] To conserve body fluids, the volume of urine produced declines and may virtually cease and, as a result, urea and potassium begin to accumulate in the blood. Blood urea may increase from normal values of 10–20 mg/100 ml to values as high as 120 mg/100 ml, and plasma potassium, although declining within the cells of the body, may increase from normal values of 5 m-equiv./1. to 6–7 m-equiv./1. and even higher at the time of death.[15,16] The loss of bicarbonate in the faeces and the accumulation of ketone bodies due to starvation results in a metabolic acidosis, and the pH of the blood falls. If the blood pH falls much below 7·2, the calf will usually succumb.

Arrythmias of the heart action occur in a large proportion of the calves that die, and it is probable that in many cases, the diarrhoea produces primary cardiac failure as a result of depletion of the potassium in the heart muscle. Such depletion has been found to occur in calves dying as a result of diarrhoea, and also in calves which have had the blood pH reduced by acid infusion.[17] Data showing the faecal losses that result from diarrhoea and of changes in the electrolyte pattern of the blood and heart muscle are given in Tables 10.1 and 10.2.

Table 10.1 AMOUNTS OF VARIOUS CONSTITUENTS EXCRETED DAILY IN THE FAECES OF CALVES WITH DIARRHOEA COMPARED WITH THOSE IN THE FAECES OF NORMAL CALVES [18]

	Normal	*Diarrhoeal*	*Ratio diarrhoeal to normal*
Water (g)	51·0	927	18·2
Dry matter (g)	12·5	93·5	7·5
Total fat (g)	4·1	37·4	9·1
Neutral fat (g)	1·5	10·6	7·0
Soaps (g)	1·9	8·3	4·4
Free fatty acids (g)	0·7	18·5	26·4
Volatile fatty acids (ml N/10 acid)	164·0	1056·0	6·4
N x 6·25 (g)	5·5	41·0	7·5
Ash (g)	1·5	10·6	7·1
Calcium (m-equiv.)	21·6	98·8	3·7
Magnesium (m-equiv.)	11·4	24·0	
Sodium (m-equiv.)	5·0	41·6	11·3
Potassium (m-equiv.)	2·2	39·9	
Phosphorus (m-equiv.)	21·0	94·0	4·4
Coliform count x 10^8	118·0	2907·0	24·6
pH	6·8	6·0	–

Table 10.2 VALUES FOR VARIOUS CHARACTERISTICS OF THE BLOOD AND MYOCARDIAL MUSCLE OF NORMAL CALVES AND THOSE WITH DIARRHOEA THAT SURVIVED OR DIED[16,17]

		Diarrhoeal	
	Normal	*Survived*	*Died*
Plasma Na (m-equiv./1.)	142	129	129
Plasma K (m-equiv./1.)	5·1	5·1	6·1
Plasma C1 (m-equiv./1.)	100	92	94
Plasma volume (ml/kg body weight)	66	59	57
Blood urea (mg/100 ml)	16	41	91
Blood pH	7·4	7·3	6·9
HCO_3 (m-mole/1.)	29	22	9
Myocardial K (mg/100 g muscle)	223	–	169

PREDISPOSING CAUSES

Although these findings indicate the sequence of events that may culminate in death, the conditions which predispose to the dominance of *E. coli* in the intestinal tract appear to be two-fold.

In infants and presumably in calves, there is a balance (eubacteriosis) between the saccharo-proteolytic organisms, including *E. coli* whose main activity is proteolytic, and the saccharolytic organisms, mainly lactobacilli, whose activity is associated with the fermentation of sugars, mainly to lactic acid. If the balance is swung too far in either direction (dysbacteriosis), the animal is likely to be affected with diarrhoea, which has been designated putrefactive in the case of the dominance of the saccharo-proteolytic organisms and fermentative when the saccharolytic organisms are dominant.[19]

The dominance of either group of organism can result from infection with highly invasive strains of the organism, from changes in the composition of the chyme passing from the abomasum, or from the use of antibiotics.

The dominance of certain strains of *E. coli*, resulting from the continual passage of susceptible calves through a calf house with a resultant build-up of 'infection', has been demonstrated. In a calf house, maintained at 12·8–15·6°C (55–60°F) and into which purchased new-born calves were introduced at intervals of every 1 to 2 days, each successive calf gained less weight than its predecessor, the incidence of diarrhoea, i.e. of faeces containing less than 10% dry matter, gradually increased and deaths eventually occurred from an *E. coli* localised intestinal infection, associated with the dominance of particular strains of *E. coli*.[4,20]

This 'build-up' of infection occurs at a greater rate if the milk substitute given after the colostrum-feeding period predisposes to the multiplication of *E. coli.* Milk substitutes are based to a large extent on skim milk powder. If, in the processing of skim milk, denaturation of the whey proteins has occurred to a marked extent, the milk will not clot properly in the abomasum and undigested protein will pass into the duodenum and allow a suitable substrate for the multiplication of *E. coli* (see p. 97).[18,21–28] Even if the whey proteins have not been denatured, too high a sodium : calcium ratio in the diet will prevent clotting, and cause diarrhoea.[29] The main harmful products of the multiplication of *E. coli* are toxins and amines, and it has been suggested that in humans the degree of injury to the intestinal walls depends on the activity of the amine oxidases and the amount of antibody present in the cell walls of the colon,[19] The feeding of very large amounts of colostrum may thus saturate the calf's tissues with antibodies and may thus protect calves from an *E. coli*, localised intestinal infection as well as from an *E. coli* septicaemia, even though the calves may be given subsequently a poor-quality milk substitute.

Thus it can be seen that putrefactive diarrhoea is probably due to abnormal amounts of protein and peptides in the chyme passing into the duodenum. This may arise from overloading the absorptive capacity of the small intestine and from disturbances in protein digestion. For instance, a diet consisting solely of skim milk, which has a very high protein content, will tend to cause diarrhoea. Overfeeding, which causes excessive curd formation in the abomasum, has been suggested as a predisposing factor to diarrhoea,[30] but recent work has indicated that very high levels of whole milk and good quality milk substitutes may be fed directly after the colostrum-feeding period without necessarily causing diarrhoea, or excessive passage of undigested protein into the duodenum.[31,32] It must also be borne in mind that there are breed differences in the digestibility of protein and that, for this reason, the extreme dairy breeds may be more susceptible to the effects of excessive protein.

Contamination of milk with *E. coli* organisms has also been suggested as a predisposing cause,[33] but whereas *E. coli* septicaemia can be produced in the colostrum-deprived calf by feeding specific strains of coli,[6] the production of an *E. coli* localised intestinal infection by feeding the organism isolated from calves that have died from this condition has usually been unsuccessful.[34]

Although supplementation of the new-born calf with vitamin A, or prepartum feeding of vitamin A to the dam has not been shown to give any protection against *E. coli* infections,[35,36] a calf born of a vitamin A-deficient dam is likely to be more susceptible to damage to the intestinal epithelial cells and thus to diarrhoea.

Outbreaks of diarrhoea associated with calves sucking their dams on particular pastures[37] may be due to the secretion in the milk of excessive amounts of non-protein nitrogenous components.

Sudden changes in diet, the use of sour milk, poor hygiene and exposure of the calf to stress in transit and to cold and damp, may all play their part in increasing the susceptibility to infection.

PREVENTION

E. coli infections are undoubtedly associated with failures on the part of farmers and dealers to exercise the rules of good animal husbandry and on the part of manufacturers to produce milk substitutes that do not predispose to this condition.

By feeding adequate quantities of colostrum to the calf during the first 24 hr of life, the incidence of *E. coli* septicaemia should be low. Antibodies against the strains of *E. coli* present in the cow's environment, associated with the globulins of colostrum, should protect the calf against death. However, new strains may be introduced with purchased calves, and the colostrum of down-calving cows or heifers purchased shortly before calving may not contain antibodies against the strains of *E. coli* on the buyer's farm. It is possible that strains present on a farm may increase in virulence as they pass through successive calves. Even though the feeding of colostrum should protect the majority of calves from an *E. coli* septicaemia, deaths from this cause still occur quite frequently under commercial conditions. These deaths often are associated with divergences from the finer aspects of good management, some of these faults are indicated below:

1. *The interval between birth and the first feed of colostrum is delayed too long.* For instance, calves born at night are assumed to have suckled and are removed from the dam the following morning and then may not be fed colostrum until the afternoon. As the absorption of antibodies declines with increasing time after birth, the calf may in fact absorb very little under such conditions. Once the meconium has been passed completely, no further absorption appears to occur.[38] Moreover, it has been suggested that the period of permeability is shorter in winter than in summer.[39]

2. *Colostrum from the first milking after calving is not given to the calf at the first feed.* This may happen if a number of cows calve at about the same time and colostrum is bulked. Colostrum, bulked from different milkings after calving will have a lower content of immune lactoglobulins than colostrum obtained from the first milking, since the high concentration of protein in colostrum, which is due largely to its content of

immune lactoglobulins, falls exponentially from 14–20% at the first milking, to 8–12% at the second, to 5–6% at the third and to 4·2–4·4% at the fourth.[40] However, it is not essential for a calf to receive colostrum from its own dam, provided that first-milking colostrum is given at the first feed.

3. *Colostrum is diluted with water, so that the concentration of immune lactoglobulins is reduced, or heated sufficiently to denature the whey proteins, including the immune lactoglobulins.* This may occur if steam is passed into colostrum or some other method is used which allows the temperature to rise above 50°C (122°F).

4. *Insufficient colostrum is offered to the calf.* Although quite small amounts of first milking colostrum should protect a calf against an *E. coli* septicaemia,[41,42] such amounts will not protect calves against an *E. coli* localised intestinal infection.[43] Thus, the greater the amount of colostrum ingested, the better it is for the calf, especially as colostrum is the main source of fat-soluble vitamins. With new-born calves left with their dams, about 7–8 kg (15–18 lb) colostrum are ingested on the first day of life, increasing to 10–12 kg (22–26 lb) by the fourth day;[44] these amounts are much larger than those usually given by bucket feeding of new-born calves removed from their dams at birth. This difference in the amount of colostrum obtained may account partly for the difference in the mortality rate of calves receiving colostrum by these two methods.

5. *Cows may be milked before parturition, with the result that the post-partum secretion is of similar composition to that of milk.*[45] To adequately protect the calves produced, colostrum should be available to them from another cow or from a deep-freeze store. Colostrum can be stored without reducing its protective value at -18 to -25°C for at least 6 months provided it is frozen in small amounts (0·6 l. ≡ 1 pt cartons). When colostrum is for any reason not available for the calf, a tetracycline antibiotic (see p. 92) should be given, both by injection and by mouth as soon after birth as possible. As an alternative, serum from a cow on the farm may be injected, or a substitute for colostrum may be fed (see Volume 1, p. 49), although this method is probably less reliable.

On some farms where *E. coli* infections appear to occur regularly, proprietary *E. coli* immune serum has been used as a preventative with good results. However, strains of coli against which the serum is ineffective may be introduced on to the farm and other treatment will be necessary. Certain sulphur drugs and antibiotics (see p. 92) have been used to good effect, but in general there is no specific cure for all outbreaks of *E. coli* infection. To enhance the value of medication with antibiotics, bacterial sensitivity tests of the organism isolated against various antibiotics may be necessary, to ascertain the most efficacious antibiotic to use.

The knowledge that the resistance by *E. coli* develops to antibiotics, both when used therapeutically, or when used prophylactically at low levels in proprietary feedstuffs, as has been the custom in the U.S.A.,[46–49] and the more recent information that resistance may be of a multiple and infective nature[50–54] and may be passed to other organisms such as *Salmonella* by contact, has led to the use of alternative methods of treatment.

The tendency for the faeces of calves with putrefactive diarrhoea to have an alkaline pH has led to the suggestion that dilute acid solutions should be given to calves to reduce the pH. With normal calves, the modal pH was about 6·3 compared with 7·3 for calves with diarrhoea. Treatment with acid solution of the calves that had diarrhoea reduced the pH of the faeces to about the normal value and seemed to cause the replacement of the *E. coli* in the faeces by lactobacilli.[55]

The possibility of electrolyte replacement therapy to rehydrate the calf suffering from diarrhoea, together with transfusion of plasma, has received increased attention lately. Electrolyte replacement treatment on its own has not been very successful. Isotonic saline (pH 6·2) (9 g sodium chloride per litre), isotonic sodium bicarbonate (pH 8·0) (12 g sodium bicarbonate per litre), or Darrow's repair solution (4 g/l. sodium chloride, 52 m-moles/l. sodium lactate, 2·7 g/l. potassium chloride)[56] are the usual electrolyte solutions suggested.[14] Mild dehydration can be overcome by oral administration, but advanced dehydration will require parenteral administration. Isotonic saline is considered by some to be unsuitable for intravenous administration as it will tend to aggravate the metabolic acidosis. The benefit of plasma in addition to electrolytes seems to lie in its protein content and thus in its buffering effect on the blood pH. For intravenous infusion, 400–800 ml plasma followed alternately by 500 ml Darrow's solution and 500 ml 5% glucose solution until a total of about 4·5 l. have been infused, has been suggested as a suitable procedure. The initial infusion should be given at a slow rate (6 ml/min) and then, after 200 ml have been infused, the rate can be increased to the maximum under gravity flow until 400–600 ml have been infused, with the remainder infused at the rate of 2–3 ml/min.[14]

FERMENTATIVE DIARRHOEA

In infants, fermentative diarrhoea, which is probably synonymous with many cases of so-called 'nutritional scours' in calves, is associated with excessive carbohydrate in the diet, or a deficiency of enzymes to degrade the carbohydrates. The only sugars that can be utilised by the young calf are glucose and lactose, but too large a quantity of these in the diet may cause diarrhoea.[18, 57-60] As mentioned earlier (see p. 6), the pre-ruminant calf cannot digest starch or the degradation products, dextrin and maltose, until at least 28 days of age, and no sucrase is present in the intestine for the hydrolysis of sucrose. Thus, the feeding of starch and its degradation products may also cause diarrhoea in the young calf.

Intakes greater than 200 g lactose and glucose per day caused the dry matter content of faeces (fat-free basis) to fall below 10%.[18] In general more than 9 g/kg live weight of 'hexose equivalent', i.e. the sum of the weight of glucose, and the weight of lactose x 1·05,[60] tends to cause 'loose' faeces in calves. However, since a high content of carbohydrate in the diet is associated generally with a low content of fat, a 'hexose equivalent' of 12 g/kg may be fed to calves after the colostrum feeding period, provided that the fat intake is about 5·5 g/kg live weight.

The predominance of the saccharolytic flora in fermentative diarrhoea results in the production of organic acids of low molecular weight, especially lactic and acetic acids, which have an irritant effect on the intestine and produce hyperperistalsis. This usually results, in the case of infants, in faeces of low pH, but sometimes fermentative diarrhoea is accompanied by secondary putrefactive diarrhoea. In general, the consequences of fermentative diarrhoea are much less severe than those of putrefactive diarrhoea.[19]

PREVENTIVE MEASURES BY THE FARMER AGAINST DIARRHOEA IN CALVES

Much can be done by the farmer to eliminate the predisposing causes of diarrhoea. The following points in particular should be borne in mind:

1. The younger the age of the calf when first given a milk substitute, the better must be the quality of the food. If given directly after the colostrum-feeding period, the milk substitute must clot readily in the abomasum. This can be tested by adding 0·2 ml commercial rennet to 100 ml of the reconstituted powder at 37°C. A good quality powder should form a firm clot within 3–4 min, a poor powder will take much longer and will produce only a flocculent curd.

2. A high-fat milk substitute is likely to give better results than a low-fat milk substitute, provided that the digestibility of the fat is high and fat globule size is small. This is especially so if the milk substitute is being fed at high levels.

3. Overfeeding of calves should be avoided, especially if their previous history when bought-in is unknown. If the calves are very thirsty on arrival, they may be offered 3–5% glucose solution. If any calf has

diarrhoea, it is probably wiser not to feed glucose, as it may aggravate the condition; water, isotonic saline or sodium bicarbonate are likely to be more suitable. When the faeces have returned to normal, 3–5% glucose may then be given, or the calf may be gradually introduced to a milk substitute.

4. Only freshly reconstituted milk, warmed to body temperature (36–38°C, 95–100°F) should be given. If it is desired to feed at a lower temperature, the quality of the milk substitute is likely to be of even greater importance.

5. Feeding should be done at regular hours, and the utensils scrubbed and sterilised after each feed.

6. Calves that are purchased under 3 weeks of age should be isolated, or alternatively, calves should not be bought until they have passed the susceptible age. If diarrhoea occurs in purchased calves, a simple blood test can be made to verify that the calves have received adequate colostrum. The routine for this test for the presence of adequate γ-globulin and associated antibodies in the sera of young calves is as follows: 5 ml solution containing 250 mg zinc sulphate ($ZnSO_4.7H_2O$)/1. is added to 0·1 ml calf serum in 1 ml distilled water. An immediate marked turbidity, followed by precipitation reaching its peak about 1 hr later, will indicate that the calf has had adequate colostrum. The solution containing the serum from a calf deprived of colostrum will remain completely clear.[61]

7. If there is a tendency for calves in a calf house to have diarrhoea, introduction of further susceptible animals should be avoided. During the winter months, the incidence of diarrhoea increases and reaches a peak in early spring. With bought-in calves, batch rearing on the 'all in–all out' method is likely to be more successful than the introduction of susceptible animals successively into the same building, a procedure that is liable to result in a 'build-up' of infection.

In many cases of diarrhoea, the duration of the attack is largely under the control of the calf rearer. During the sixth to twelfth days of life, the calf appears to be particularly susceptible. By careful observation, it is possible to anticipate the onset of diarrhoea on the day before it occurs, and milk can then be withheld or reduced in quantity with the result that the calf recovers quickly.

The following signs of impending diarrhoea should be looked for: dry muzzle, thick mucous appearing from the nostrils, very firm dung, refusal of milk, a tendency to be prostrated and a high rectal temperature (over 39·3°C (102·8°F)).

Various methods have been suggested for the treatment of diarrhoea in calves by the farmer and two of them are given below:

1. At the first onset of diarrhoea, warm water only or warm water

containing 0·9% salt (9g sodium chloride per litre (⅓ oz salt/2 pt water)) should be fed for 24 hr at the rate of 1·1 kg (2 pt) per feed in three feeds per day. During this period no other dry or liquid feed should be given. For the next 24 hr 1·7 kg (3 pt) of milk with 2·6 kg (4½ pt) of water should be given, and for the following 24 hr feed 2·6 kg of milk with 1·7 kg of water in each case in 3 feeds per day. If the calf is now normal, undiluted milk may then be offered.

2. This method is simpler than the preceding one. It consists of starving the calf for one feed and then reducing the daily ration of milk to that required for maintenance of body weight under normal conditions (see Table 2.1); for a 27 kg (60 lb) calf, this is about 2·3 kg (4 pt) per day, for an 36 kg (80 lb) calf this is about 2·8 kg (5 pt) per day and for a 45 kg (100 lb) calf 3·4 kg (6 pt) per day. The calf should be maintained on this ration, given in two or three feeds, until the faeces show signs of becoming firmer. The quantity of milk may then be gradually increased.

SALMONELLOSIS

Infection in calves is usually by the organisms *Salmonella dublin* or *Salmonella typhi-murium*, with infection by the former organism being more common.[62,63] Between 1958 and 1960, 64% of cases of Salmonellosis diagnosed at Veterinary Investigation Centres were associated with *S. dublin* and 32% with *S. typhi-murium.* Of the post-mortem examinations made on calves up to 6 months of age during the period 1959–1961, 16% were associated with a salmonella septicaemia and 7% with a salmonella gastro-enteritis. Salmonellosis was responsible for 6% of the total deaths in home-bred calves and 30% in purchased calves.[3] A more recent report has shown a ratio of *S. typhi-murium* to *S. dublin* infections in cattle (mainly calves) as 1 : 3 in the years 1961–1964, with a marked increase in the incidence of *S. typhi-murium* in 1965, so that the ratio increased to 3 : 4.[64]

S. dublin is mainly a pathogen of cattle, although it may infect man, and in certain areas, such as South Wales and Somerset, may be endemic, so that ½–1% of adult cattle are excreting the organism in their faeces. These areas are often of rough grazing where cattle are drinking from streams or dykes, and where liver fluke infection is also common. On the other hand, *S. typhi-murium* is able to infect practically all species, including man, and does not appear to be endemic in particular areas.

The two infections produce similar clinical symptoms, which are variable, and the calf may become infected as early as the fourth day of life, although the incidence is highest during the period 1–4 weeks of age. In acute cases a septicaemia may occur, but in milder forms diarrhoea, often blood-stained, may be the only symptom and this may be transient.

Calves affected more severely become emaciated although appetite may not be completely lost. Some of these calves make a slow recovery, but others show a short period of weakness and prostration before death. Calves that recover do not usually continue to excrete the organism in their faeces. Sometimes the disease is accompanied by lameness, associated with arthritis, and often the calves are also infected with 'virus pneumonia' (see p. 133).

Prevention lies in the identification and removal of the source of infection. An adult cow may be a carrier of the disease, having recovered from an infection which may have been so mild that no symptoms were present. Pregnant cows that either develop clinical salmonellosis or become active carriers, may abort their calves, or if the calves are carried to full-term they may be infected *in utero*. Such calves may die shortly after birth. If not infected *in utero*, the calves are likely to acquire the infection from the carrier dam soon after birth. Such calves do not usually die, presumably because of the protection afforded by the dam's colostrum, but they may infect other calves whose illness may be much more severe. With *S. dublin* infections, a carrier cow may continue to pass large numbers of organisms in its faeces for many years, but with *S. typhi-murium*, the carrier state is usually of shorter duration. Moreover organisms may be present in an internal organ, such as a lymph gland, and may not be excreted in the faeces until the animal is exposed to stress. Milk from a clinically infected cow may be the cause of infection but milk from carrier cows is seldom infected unless contaminated by faeces. Identification of a carrier cow is possible by laboratory examination of the sera and faeces of suspected cows, although sometimes the organisms are only excreted intermittently.

In endemic areas, which are mainly breeding areas, the outbreaks in calves are often quite mild and the mortality sporadic, but in purchased calves, including those bought-in from endemic areas, which have been exposed to further infection in markets and transport, to chilling and to changes in diet, the disease may be serious with a high mortality rate. The development of the early-weaning system (see Volume 1, p. 80), in which the energy intake is low during the most susceptible period of life, may also be responsible for the increased incidence of the disease. Other sources of infection are the use of infected calf foods, the presence of infected rats and mice in the precincts of the calfhouse, or a human carrier.

Purchased calves should be segregated from older batches, and should be carefully looked after on arrival, so that any infection may be quickly diagnosed and recovery from the stress of the journey expedited. In particular, they should be kept warm, and, if there is any sign of looseness of faeces, they should forego a feed, but have access to water. There is considerable diversity of opinion as to whether 3–5% glucose solution should be given to the calves on arrival. If the journey has been short and the calves recently fed, additional carbohydrate may predispose to fermentative diarrhoea (see p. 127), especially if the milk substitute that had been used contained more carbohydrate than is present in whole milk. On the other hand, if the calves have been without food for a considerable length of time, then the feeding of glucose, as a readily available source of energy, can be recommended.

A live vaccine (*S. dublin* strain 51) is now available, which gives a good degree of immunity against *S. dublin* and should be injected subcutaneously at least one week before the calves leave the premises on which they were born.[65] This vaccine also gives a degree of cross-immunity to *S. typhi-murium*, as two of the seven 'O' antigens which occur are common to both organisms.[66] Buyers of calves should ensure that the calves have been vaccinated and have come from herds free from adult carriers. Even if vaccination is delayed until calves arrive on the premises of the rearer, with the disadvantage that the calves by this time may already be infected, vaccination still has some protective value.[67] As antibiotics or other antibacterial drugs, if given after vaccination may reduce the immune response, only such drugs given orally, which are not readily absorbed from the intestine, should be used.

If an outbreak on a farm is diagnosed, sick animals should be segregated. On no account should children and dogs have access to these animals, owing to the risk of infection, and the attendant should take precautions, by using protective clothing, to prevent the spread of disease to other stock or into his home.

After an outbreak, the building must be thoroughly cleaned, disinfected and fumigated with formalin or a proprietary aerosol, as the organism may survive for up to 6–12 months in dust as well as in faeces.

NAVEL-ILL OR JOINT-ILL

In this disease, the infective agent appears to enter via the navel. The mortality rate is high, and in acute cases death is rapid with no specific symptoms. In less acute cases there is usually some swelling of the navel with abscess formation, or the infection may spread to the liver with more serious effects. In addition, the calf loses its appetite, has a slightly elevated temperature and tends to be prostrated. The joints, especially the hocks and knees, usually become swollen and painful.

A number of different bacteria are associated with this condition, and recently an organism of the psittacosis-lymphogranuloma venereum group (PLV) and a species of Mycoplasma have been associated with polyarthritis in calves.[68,69]

Where navel disease is prevalent on a farm, all navel cords should be dressed at birth (for details see Volume 1, p. 43) and the buildings used for calving should be thoroughly disinfected.

CALF DIPHTHERIA

This disease, caused by the organism *Fusiformis necrophorus*, attacks calves usually before they are 6 weeks old, and may occur as early as 4 days from birth. It is characterised by the formation of ulcers on the mucous lining of the mouth and pharynx. The base or sides of the tongue, the inside of the cheeks adjacent to the teeth and the lips are infected first, and the lesions may then spread throughout the mouth and nasal passages, and down the trachea to the lungs. The destroyed tissue is greyish-yellow in colour, and around the edge of an ulcer the mucous lining is slightly raised, reddened and granulated.

The infected calf will tend to slobber and swellings may be seen at the side of the cheek or in the throat region. Where the lesions involve the nasal and respiratory passages, laboured breathing, coughing and a sticky, greenish-yellow discharge from the nostrils will be apparent. The swollen tongue may also protrude from the mouth and an offensive odour is often emitted. In addition, the calf loses its appetite, goes back in condition and may show signs of diarrhoea. The mortality rate is usually high, acute cases dying within a week, whereas a calf suffering from a milder attack may survive for a number of weeks.

When the disease has been diagnosed on a farm, the sick animals should be segregated, and the calf pens and utensils thoroughly disinfected. All healthy animals should be examined each day for the early detection of new cases. Veterinary advice should be obtained for the treatment of the sick animals.

CALF PNEUMONIA (ENZOOTIC BRONCHITIS)

With the development of intensive systems of production there has been an increased incidence of so-called 'virus pneumonia' in calves. The cause of the disease is extremely complex but appears to be the result of the interaction of one or more viruses with several predisposing causes, such as the stress of being moved long distances, or through markets, producing the so-called Transit Fever and Shipping Fever in the U.S.A., and the physical environment in which calves are housed. The infection by viral agents is often followed by a secondary bacterial invasion, usually by *Pasteurella haemolytica* or *Corynebacterium pyogenes.* Calves exposed to parainfluenza type 3 virus followed by pasteurella organisms 24 hr later show more marked signs of transit fever than those exposed to pasteurella alone.[70-72]

In general, the outbreaks, which occur mainly in the winter months, do not affect calves younger than 1 month of age, suggesting that the calves are probably protected during this time by the passive immunity received from their dams. Whether the protection afforded is to the secondary invaders or to the virus, or both, is not clear. The incidence of infection (morbidity) is usually high, but the mortality rate may be very variable. It is usually low, but on occasions may be severe; the severity of the mortality may be associated with the degree of secondary bacterial invasion that has occurred.

The symptoms are variable, but there is usually a nasal discharge, which is sometimes thin and watery, but at other times may be thick and purulent. A slight discharge may also occur from the eyes. In most outbreaks, there is a dry cough, which often persists after the calf has recovered, and is particularly noticeable on exercise. Lesions of variable extent usually occur in the lungs, especially in the apical lobes (see Plate 10.2). If obvious signs of respiratory distress are present, the calf will invariably have a high temperature, but not necessarily in more chronic cases. Diarrhoea is also often associated with the condition.

The method adopted to ascertain whether calves have been clinically infected with a particular virus is by the paired serum sample method, in which the first sample is taken at the first signs of the disease and the second three weeks later. If there is a four-fold increase in antibody titre, it is considered that the calf was infected with that particular virus at the time of the first sample.[73]

In Great Britain four viruses have been shown to cause pneumonia in calves. They are the infectious bovine rhinotracheitis virus (IBR), parainfluenza type 3 (PI3), reovirus and adenovirus. In addition the psittacosis-lymphogranulonia venereum (PLV) group of organisms have also been implicated. Other viruses, including bovine viral diarrhoea virus (BVD) (see p. 137) have been isolated from calves with respiratory infection.[74]

The P13 virus, widespread throughout the world, and adeno- and reovirus appear to be the most important viral agents associated with pneumonia in calves in Great Britain. At first sight, it would appear that the PI3 virus is of the most importance, as 84% of serum samples obtained from animals at slaughter-houses had titres greater than 1 : 32[75] and 23% of serum samples had antibodies against adenovirus.[76] However, using the paired sample method, 24% of 150 outbreaks of respiratory

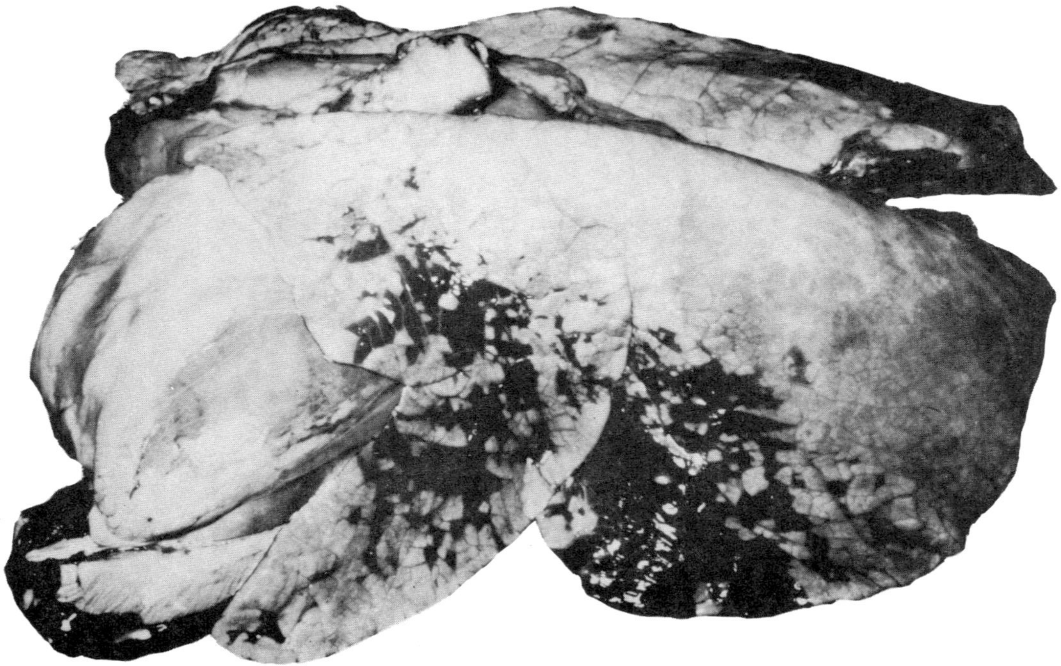

Plate 10.2. Left lateral aspect of the lungs, showing lesions in calf infected with parainfluenza type 3. (Dawson, P. S., Derbyshire, J. H., and Lamont, P. H. Res vet. Sci. **6**, *108 (1965)) (By permission of the Controller of Her Majesty's Stationery Office. Crown Copyright)*

infection in calves was associated with PI3, 29% of 133 outbreaks with adenovirus and 26% of 111 outbreaks with reovirus.[77] IBR virus seems to be of much less importance, antibody titres being found only in 2% of serum samples obtained from animals at slaughter-houses[75] and only one out of 108 outbreaks of respiratory disease could be associated with the virus. However, both this virus and that of bovine viral diarrhoea have an affinity for lymphoid tissue, and failure to produce antibodies may be the result of destruction of the immunologically competent cells.[78,79]

Differential diagnosis of the viral agent involved from the lesions in the lungs is difficult in natural outbreaks, because of the much greater effect of the secondary bacterial invaders. However, it appears that a watery discharge and conjunctivitis are associated with the IBR virus and that diarrhoea accompanying the pneumonia most frequently occurs with adenovirus, IBR virus and PLV organisms.

It has been suggested in the U.S.A. that adenoviruses associated with diseases in calves may be of public health importance, since two strains, antigenically related to human adenoviruses and isolated from the faeces of apparently healthy cows, caused respiratory illness in calves.[80]

Treatment of virus pneumonia is largely aimed at reducing the effect of secondary bacterial invasion. Although poor ventilation and high humidity of calf houses have been implicated as predisposing causes in outbreaks of the disease (see Volume 1, p. 137), at the moment the relative importance of various factors, such as air temperature, air movement, relative humidity, ammonia concentration, and cubic area per calf has not been elucidated.

INFECTIOUS BOVINE RHINOTRACHEITIS (IBR)

Besides causing enteritis, the IBR virus, which is related to the herpes virus in man, causes necrosis of the mucosa of the upper respiratory tract rather than of the lower tract, although pneumonia may result from a secondary infection. It also causes necrosis of the lymphoid tissue and of the adrenal cortex[78] and may affect the genital tract causing infectious pustular vulvovaginitis.

The first symptom of the disease is the intense reddening of the conjunctiva and mucous membranes of the nose. The discharge from the eyes and nose is clear at first but gradually becomes thick and purulent. The temperature may rise to a high level and the condition may be accompanied by a moist cough.

In the U.S.A. the IBR virus is considered as one of the inciting agents of shipping fever and a vaccine against it is available.

MUCOSAL DISEASE

Mucosal disease is caused by the bovine viral diarrhoea virus (BVD)–mucosal disease virus complex. It appears to be quite widespread in Great Britain, as 45% of serum samples obtained from animals at slaughter-houses contained antibodies to the condition.[81]

The disease is associated with ulcerations of the mucosa, in particular of the oral cavity and the abomasum, and in enteritis resulting in diarrhoea.

In the first stage the calf has a high temperature and no apparent symptoms; the temperature usually recedes for about 3 days and then returns to a much higher level with ulcerations in the mouth and reddening of the nostrils and oral mucosa especially around the teeth. There is also a mucous nasal discharge and inappetence, and sometimes increased lachrymation.[82] In mild cases these are the only symptoms, but in more severe cases, diarrhoea is marked and the faeces may become blood stained. The ulcerations are severe and besides occurring on the muzzle, lips, and mouth, may also occur between the claws of the hoof. The nasal discharge is also profuse and in the early stages there is much evidence of thirst. Abortions have been associated with some outbreaks of the disease.

Like all virus diseases of calves, the condition does not usually affect colostrum-fed calves until they are over 1 month of age, and the condition seems to be more severe when it occurs in self-contained herds, probably due to the low level of herd-immunity.[83]

It has been postulated that, as the ulcerations in the mouth are similar to those of hyperkeratosis, vitamin A deficiency may possibly predispose calves to the condition.[84] Lesions in the abomasum and in the lymphoid tissue of the intestines are characteristic of mucosal disease, and it has been suggested that failure to produce antibodies against the disease may be due to the destruction of lymphoid tissue. There is also a suggestion of genetic susceptibility to the disease, as in one outbreak only calves from one sire were affected.[79]

In mild cases of the disease, calves recover quickly, but in severe cases, the mortality rate may be high. Simultaneous infection with Infectious Bovine Rhinotracheitis and Mucosal Diseases has been shown to produce more severe effects than with either virus singly.[78] As there is no specific treatment for the disease, the aim is to minimise the effect of any secondary bacterial invasion.

BLACK QUARTER (QUARTER-ILL OR BLACKLEG)

Blackleg, caused by the organism *Clostridium chauvei*, has in exceptional cases been responsible for the death of calves under 2 months of age, which have been kept in pens since birth. However, it is more usual for calves to be affected when sucking their dams out at grass, in those areas of the country where the incidence of blackleg is high. Low-lying damp permanent pastures seem to be associated with the disease, which is contracted by animals eating food infected with clostridial spores or by infection through a wound.

The affected animal has a fever and shows inappetence, followed by the appearance of swellings in the muscles of the body, which are at first hot and painful, but later become cold and painless. The skin over the affected muscles becomes hard and stiff, and a crackling noise occurs when pressure is applied, owing to the stiffness of the skin, and to the gas produced by the bacteria, beneath the skin. The organism produces a toxin, which will rapidly lead to death. If an animal dies from the disease, the carcase should be burnt, deeply buried or removed from the field intact, as otherwise it may be a source of infection due to release of bacteria and spores.

Treatment with antibiotics and serum may be effective in early cases. As a preventive measure, calves on farms where blackleg occurs, are usually immunised when 4–12 weeks of age or a few weeks prior to turning out to grass.

BLOAT

Bloat occurs as a result of excessive gas production either in the abomasum or the rumen of calves.

Abomasal bloat occurs in some calves immediately after feeding a milk substitute or even whole milk. It seems to be associated with the proliferation of an undesirable or excessive flora in the abomasum, possibly of *E. coli* that have spread to the abomasum after having multiplied in the duodenum as a result of excessive undigested protein passing from the abomasum, but more probably of lactobacilli as a result of too high a level of soluble carbohydrate in the diet. Certainly, the feeding of milk substitutes of low fat content seem to be one of the predisposing causes. Treatment lies mainly in reducing milk intake, and thereafter using a diet that does not predispose to the condition. Release of gas can be affected by means of a stomach tube, but the process will usually be necessary after each meal. Bloat of milk-fed calves is also associated with muscular dystrophy (see p. 83).[85]

Ruminal bloat does not usually occur in calves fed conventional rations, but is restricted to regimes where roughage is not included in the diet, particularly where all-concentrate diets are given in restricted amounts rather than *ad lib.*[86] and where the fibrous characteristics of the concentrate mixture are lacking. Thus diets containing large amounts of ground maize are more likely to cause bloat than crushed barley.[87]

RICKETS

This disease of the young calf is characterised by a failure of the bones to calcify normally.

As mentioned earlier, the cause of the condition is a deficiency of vitamin D, or faulty proportions or insufficient amounts of calcium and phosphorus. For the prevention of rickets, the reader should refer to minerals (p. 38) and vitamins (p. 77).

HYPOMAGNESAEMIA

This condition is associated with a fall in the magnesium content of the blood (see p. 41). Affected calves show symptoms similar to those of cows suffering from lactation tetany.[88] The calf shows much excitement, the head is retracted, the whites of the eyes are very apparent, owing to an upward rotation of the eyeballs, and there is often continual movement of the ears. Walking is uncoordinated and the calf tends to exaggerate the lifting of its feet. When touched the animal seems to be hypersensitive. In the final stages convulsions occur; the animal falls on its side with neck stretched out, the jaws clenched and all respiration movements stopped. Control of urination and defaecation is lost and the eyes alternately bulge and sink in their sockets.

The best criterion for confirming that a calf has died of hypomagnesaemia is given by chemical examination of the bone. Ratios of calcium to magnesium in the bones of affected animals are 90 : 1 or greater, compared with 60 : 1 or less in normal calves.

In treating this condition, it is necessary to restore the depleted stores of magnesium in the bone as well as to increase the blood magnesium. The blood magnesium content can be restored by intravenous injection of about 100 ml of a solution containing 14 g(½ oz) calcium borogluconate and 3·6 ml (1 dr) magnesium lactate followed by 14 g (½ oz) magnesium sulphate in 50 ml of water given by subcutaneous injection. To restore the bone magnesium it is necessary to feed 14 g (½ oz) magnesium oxide per day in the milk or concentrate mixture for several weeks.

Prevention of the condition can be achieved by feeding hay and concentrates or by the addition of 5 g magnesium oxide or 8 g magnesium carbonate per day. In one experiment,[89] 0·5 g magnesium as the carbonate, i.e. 1·75 g magnesium carbonate given daily largely prevented tetany in calves given large quantities of milk, but 2 g magnesium carbonate per day did not appear to be enough for animals over 2 months of age.

CEREBRO-CORTICAL NECROSIS (see pp. 84–85)

MUSCULAR DYSTROPHY (see p. 81)

LEAD POISONING

Cattle, and especially calves, seem to be particularly susceptible to lead poisoning. The commonest sources of lead are from painted woodwork and metalwork, discarded paint tins, old lead plates from batteries, painted tarpaulins and vegetation in orchards recently sprayed with lead arsenate solution.

The lethal dose for young calves is about 156–250 mg/kg body weight ($^1/_4$ to $^2/_5$ oz/100 lb body weight), but older calves may survive doses of this order. Calves tolerated 6 mg lead per kilogramme body weight for a period of 3 yr, but a single dose of 200–400 mg/kg may produce symptoms within 24 hr.[90, 91] However, a time lag of several days may occur before the onset of symptoms, the animal then dying within a few hours. A healthy calf may be suddenly found dead, or it may have a fit from which it does not recover. In less acute cases of lead poisoning, the calf is dull, dejected and has no appetite. Following this the eyes become sunken, the coat staring, and there is evidence of abdominal pain and teeth grinding with salivation. There is often severe constipation and cessation of urination, and in some cases the constipation is followed by diarrhoea. In the next phase, which may be the first in acute cases, the calf becomes greatly excited, attempting to push against the walls of its pen, appears to be blind and staggers around with rolling eyes and frothing mouth. After collapse, muscular spasms and tetany occur before death.

In acute cases there is no treatment that is likely to be entirely effective, but in sub-acute cases a purgative should be used. Calcium versenate* given daily for two periods of 6 days, as a 25% solution injected subcutaneously or intravenously, together with saline solution to reduce dehydration, has been partially successfully in the cure of calves exhibiting the nervous symptoms resulting from lead poisoning.[92] The lead becomes complexed with the EDTA* and is eliminated in the bile and urine without apparently causing damage to the kidneys. Magnesium

*The commercial name of the calcium disodium salt of ethylenediamine tetra-acetic acid (Ca EDTA).

sulphate (Epsom salts) may also be of particular benefit by forming insoluble lead sulphate in the alimentary tract of the calf.

REFERENCES

1. SOJKA, W. J. '*Escherichia coli* in domestic animals and poultry' *Rev. Ser. Commonw. Bur. Anim. Hlth* no. 7 (1965)
2. LOVELL, R. *Vet. Revs. Annot.* **1**, 1 (1955)
3. Veterinary Investigation Service *Vet. Rec.* **76**, 1139 (1964)
4. WOOD, P. C. *J. Path. Bact.* **70**, 179 (1955)
5. BRIGGS, C., LOVELL, R., ASCHAFFENBURG, R., BARTLETT, S., KON, S. K., ROY, J. H. B., THOMPSON, S. Y. and WALKER, D. M. *Br. J. Nutr.* **5**, 356 (1951)
6. FEY, H. *Schweiz. Arch. Tierheilk*, **104**, 1 (1962)
7. FEY, H. and MARGADANT, A. *Zentbl. VetMed.* **9**, 767 (1962)
8. FEY, H., LANZ, E., MARGADANT, A. and NICOLET, J. *Dt. tierärztl. Wschr.* **69**, 581 (1962)
9. INGRAM, P. L. *Symp. Soc. gen. Microbiol.* no. **14**, 122 (1964)
10. INGRAM, P. L., LOVELL, R., WOOD, P. C., ASCHAFFENBURG, R., BARTLETT, S., KON, S. K., PALMER, J., ROY, J. H. B. and SHILLAM, K. W. G. *J Path. Bact.* **72**, 561 (1956)
11. SMITH, T. and ORCUTT, M. L. *J. exp. Med.* **41**, 89 (1925)
12. SMITH, H. W. *J. Path. Bact.* **90**, 495 (1965)
13. INGRAM, P. L. 'Observations on the pathology and pathogenesis of experimental colibacillosis in calves' *Ph.D. Thesis* (University of London, 1962)
14. WATT, J. G. *Vet. Rec.* **77**, 1474 (1965)
15. ROY, J. H. B., SHILLAM, K. W. G., HAWKINS, G. M., LANG, J. M. and INGRAM, P. L. *Br. J. Nutr.* **13**, 219 (1959)
16. FISHER, E. W. *Br. vet. J.* **121**, 132 (1965)
17. FISHER, E. W. and McEWAN, A. D. *Br. vet. J.* **123**, 4 (1967)
18. BLAXTER, K. L. and WOOD, W. A. *Vet. Rec.* **65**, 889 (1953)
19. WEIJERS, H. A. and van de KAMER, J. H. *Ntr. Abstr. Rev.* **35**, 591 (1965)
20. ROY, J. H. B., PALMER, J., SHILLAM, K. W. G., INGRAM, P. L. and WOOD, P. C. *Br. J. Nutr.* **9**, 11 (1955)
21. SHILLAM, K. W. G. and ROY, J. H. B. *Int. Congr. Anim. Prod. VII, Hamburg* **3**, 276 (1961)
22. SHILLAM, K. W. G., DAWSON, D. A. and ROY, J. H. B. *Br. J. Nutr.* **14**, 403 (1960)
23. SHILLAM, K. W. G., ROY, J. H. B. and INGRAM, P. L. *Br. J. Nutr.* **16**, 267 (1962)
24. SHILLAM, K. W. G., ROY, J. H. B. and INGRAM, P. L. *Br. J. Nutr.* **16**, 585 (1962)
25. SHILLAM, K. W. G., ROY, J. H. B. and INGRAM, P. L. *Br. J. Nutr.* **16**, 593 (1962)
26. SHILLAM, K. W. G. and ROY, J. H. B. *Br. J. Nutr.* **17**, 171 (1963)
27. SHILLAM, K. W. G. and ROY, J. H. B. *Br. J. Nutr.* **17**, 183 (1963)
28. SHILLAM, K. W. G. and ROY, J. H. B. *Br. J. Nutr.* **17**, 193 (1963)
29. KASTELIC, J., BENTLEY, O. G. and PHILLIPS, P. H. *J. Dairy Sci.* **33**, 725 (1950)
30. SHEEHY, E. J. *Scient. Proc. R. Dubl. Soc.* **21**, 73 (1934)
31. MYLREA, P. J. *Res. vet. Sci.* **7**, 407 (1966)

32. ROY, J. H. B., GASTON, H. J., SHILLAM, K. W. G., THOMPSON, S. Y., STOBO, I. J. F. and GREATOREX, J. C. *Br. J. Nutr.* **18**, 467 (1964)
33. THOMSON, S. *J. Hyg., Camb.* **54**, 311 (1956)
34. ROY, J. H. B. *Vet. Rec.* **76**, 511 (1964)
35. ASCHAFFENBURG, R., BARTLETT, S., KON, S. K., ROY, J. H. B., SEARS, H. J., THOMPSON, S. Y., INGRAM, P. L., LOVELL, R. and WOOD, P. C. *Br. J. Nutr.* **7**, 275 (1953)
36. MURPHY, T. *J. Dep. Agric. Repub. Ire.* **50**, 5 (1953–54)
37. SHANKS, P. L. *Vet. Rec.* **62**, 315 (1950)
38. ROY, J. H. B. 'Studies in calf nutrition with specific reference to the protective action of colostrum' *Ph.D. Thesis* (University of Reading, 1956)
39. FISHER, E. W. *Vet. Rec.* **77**, 1482 (1965)
40. PARRISH, D. B., WISE, G. H., HUGHES, J. S. and ATKESON, F. W. *J. Dairy Sci.* **33**, 457 (1950)
41. ASCHAFFENBURG, R., BARTLETT, S., KON, S. K., WALKER, D. M., BRIGGS, C., COTCHIN, E. and LOVELL, R. *Br. J. Nutr.* **3**, 196 (1949)
42. ASCHAFFENBURG, R., BARTLETT, S., KON, S. K., ROY, J. H. B., WALKER, D. M., BRIGGS, C. and LOVELL, R. *Br. J. Nutr.* **5**, 171 (1951)
43. LANG, J. M., ROY, J. H. B., SHILLAM, K. W. G. and INGRAM, P. L. *Br. J. Nutr.* **13**, 463 (1959)
44. WALKER, D. M. *Bull. Anim. Behav.* **8**, 5 (1950)
45. ASCHAFFENBURG, R., BARTLETT, S., KON, S. K., ROY, J. H. B., WALKER, D. M., BRIGGS, C. and LOVELL, R. *Br. J. Nutr.* **5**, 343 (1951)
46. SMITH, H. W. and CRABB, W. E. *Vet. Rec.* **68**, 274 (1956)
47. SMITH, H. W. and CRABB, W. E. *J. gen. Microbiol.* **15**, 556 (1956)
48. INGRAM, P. L., SHILLAM, K. W. G., HAWKINS, C. M. and ROY, J. H. B. *Br. J. Nutr.* **12**, 203 (1958)
49. GLANTZ, P. J. *Cornell Vet.* **52**, 552 (1962)
50. OCHIAI, K., YAMANAKA, T., KIMURA, K. and SAWADA, O. *Nippon Iji Shimpo,* no. 1861, 34 (1959)
51. AKIBA, T., KOYAMA, K., ISHIKI, Y., KIMURA, S. and FUKUSHIMA, T. *Jap. J. Microbiol.* **4**, 219 (1960)
52. ANDERSON, E. S. *Br. med. J. ii,* 1289 (1965)
53. SMITH, H. W. *Vet. Rec.* **80**, 464 (1967)
54. WALTON, J. R. *Lancet ii,* 1300 (1966)
55. COWIE, R. S. *Vet. Rec.* **76**, 1516 (1964)
56. Medical Research Council, 'The treatment of acute dehydration in infants' *Med. Res. Coun. Memo.* no. 26. (H.M.S.O., London, 1952)
57. ROJAS, J., SCHWEIGERT, B. S. and RUPEL, I. W. *J. Dairy Sci.* **31**, 81 (1948)
58. FLIPSE, R. J., HUFFMAN, C. F., WEBSTER, H. D. and DUNCAN, C. W. *J. Dairy Sci.* **33**, 548 (1950)
59. CUNNINGHAM, H. M. and BRISSON, G. J. *Can. J. Anim. Sci.* **37**, 152 (1957)
60. WALKER, D. M. and FAICHNEY, G. J. *Br. J. Nutr.* **18**, 209 (1964)
61. ASCHAFFENBURG, R. *Br. J. Nutr.* **3**, 200 (1949)
62. GIBSON, E. A. *Vet. Rec.* **73**, 1284 (1961)
63. GIBSON, E. A. *Agriculture, Lond.* **73**, 213 (1966)
64. STEVENS, A. J., GIBSON, E. A. and HUGHES, L. E. *Vet. Rec.* **80**, 154 (1967)
65. SMITH, H. W. *J. Hyg., Camb.* **63**, 117 (1965)
66. RANKIN, J. D., NEWMAN, G. and TAYLOR, R. J. *Vet. Rec.* **78**, 765 (1966)
67. RANKIN, J. D., TAYLOR, R. J. and NEWMAN, G. *Vet. Rec.* **80**, 720 (1967)
68. HUGHES, K. L., EDWARDS, M. J., HARTLEY, W. J. and MURPHY, S. *Vet. Rec.* **78**, 276 (1966)

69. STORZ, J., SHUPE, J. L., SMART, R. A. and THORNLEY, R. W. *Am. J. vet. Res.* **27**, 987 (1966)
70. HEDDLESTON, K. L., REISINGER, R. C. and WATKO, L. P. *Am. J. vet. Res.* **23**, 548 (1962)
71. HETRICK, F. M., CHANG, S. C., BYRNE, R. J. and HANSEN, P. A. *Am. J. vet. Res.* **24**, 939 (1963)
72. SORENSEN, D. K., JOHNSON, D. W. and HOYT, H. H. *Nord. VetMed* **16**, Suppl. 1, 543 (1964)
73. HARBOURNE, J. F. *Vet. Rec.* **78**, 749 (1966)
74. OMAR, A. R. *Vet. Bull.* **36**, 259 (1966)
75. DAWSON, P. S. and DARBYSHIRE, J. H. *Vet. Rec.* **76**, 111 (1964)
76. DARBYSHIRE, J. H. and PEREIRA, H. G. *Nature, Lond.* **201**, 895 (1964)
77. DAWSON, P. S., LAMONT, P. H. and DARBYSHIRE, J. H. In the press (1966)
78. TYLER, D. E. and RAMSEY, F. K. *Am. J. vet. Res.* **26**, 903 (1965)
79. SHOPE, R. E. Jr, *Diss. Abstr.* **26**, 634 (1965)
80. MOHANTY, S. B. and LILLIE, M. G. *Proc. Soc. exp. Biol. Med.* **120**, 679 (1965)
81. PATERSON, A. B. *Vet. Rec.* **74**, 1384 (1962)
82. BAKER, J. A., YORK, C. J., GILLESPIE, H. J. and MITCHELL, G. B. *Am. J. vet. Res.* **15**, 525 (1954)
83. HUCK, R. A. *Vet. Rec.* **69**, 1207 (1957)
84. LOOSMORE, R. M. *Vet. Rec.* **76**, 1335 (1964)
85. BLAXTER, K. L., WOOD, W. A. and MACDONALD, A. M. *Br. J. Nutr.* **7**, 34 (1953)
86. WHITELAW, F. G., PRESTON, T. R. and MACLEOD, N. A. *Anim. Prod.* **5**, 227 (1963)
87. PRESTON, T. R., AITKEN, J. N., WHITELAW, F. G., MacDEARMID, A., PHILIP, E. B. and MACLEOD, N. A. *Anim. Prod.* **5**, 245 (1963)
88. BLAXTER, K. L., ROOK, J. A. F. and MACDONALD, A. M. *J. Comp. Path. Ther.* **64**, 157 (1954)
89. BLAXTER, K. L. and SHARMAN, G. A. M. *Vet. Rec.* **67**, 108 (1955)
90. ALLCROFT, R. *J. comp. Path. Ther.* **60**, 190 (1950)
91. ALLCROFT, R. and BLAXTER, K. L. *J. comp. Path. Ther.* **60**, 209 (1950)
92. HOLM, L. W., RHODE, E. A., WHEAT, J. D. and FIRCH, G. *J. Am. vet. med. Ass.* **123**, 528 (1953)

CHAPTER ELEVEN

Parasitic Infections of Calves

COCCIDIOSIS

Coccidiosis is caused by a protozoan parasite belonging to the genus *Eimeria*, which is specific to cattle. The coccidia, of which there are a number of different species, form spores which are capable of surviving on the ground for many months. Cattle harbouring this parasite produce large numbers of the spores, or 'oocysts', which are passed out in the faeces and, after a minimum ripening period of 2 days under conditions of warmth and moisture, are capable of infecting other animals.

Calves up to 1 year of age are most susceptible, and the disease may become widespread among a group. Calves become infected by swallowing infective oocysts with their food or drinking water. On entering the digestive tract, the spores develop into active coccidia which penetrate the walls of the intestine, and multiply, damaging the mucosal lining. In due course, oocysts are produced and released in the faeces.

Coccidiosis is most common in marshy districts during the summer and early autumn, although acute outbreaks may occur as early as 4–8 weeks of age in housed stock.[1] Most healthy calves are infected with a few coccidia but show no clinical symptoms.

Affected animals often have foul smelling diarrhoea containing blood either as liquid or in clots, and stand straining with backs arched in a characteristic posture for several minutes. The animals gradually become anaemic, weak and emaciated. The disease frequently occurs where a number of young animals are crowded together. If an epidemic breaks out, the infection becomes less severe as the epidemic progresses, possibly due to the development of an immunity in the affected calves. Adult cattle are resistant, but may act as carriers and infect susceptible young calves.

Young calves should never be introduced into crowded quarters occupied by a group of calves of various ages. As most animals carry coccidia, the pens occupied by the original group of calves will carry a substantial number of infective oocysts available for reinfection and, ultimately, large numbers of oocysts will be discharged by most of the animals. Calves subsequently introduced are exposed to a heavy infestation of oocysts, and the process is repeated until the parasite population is raised to a dangerous level.

Prevention of reinfection is essential. This can be achieved by cleaning out the building every two days, as this period is insufficient for the oocysts to become infective. Alternatively, the calves should be tied up in stalls, so that they cannot take food or water contaminated by faeces. Attempts to inoculate calves against the condition, including the use of irradiated oocysts has not been successful.[2]

Affected calves should be isolated, treated by a veterinary surgeon and, after treatment, should be removed to clean quarters. Bedding used by affected cattle should be carried right away from the building and should be buried deeply in a manure heap. Ammonia and other products of decomposition will effectively destroy the oocysts. The walls of the pens should be washed with hot soda solution, followed by rinsing and spraying with a 10% solution of ammonia.

PARASITIC GASTRO-ENTERITIS

This disease is caused by a heavy infestation of the abomasum or small intestine with small round worms (nematodes) belonging to the sub-family Trichostrongylidea. *Ostertagia ostertagi* is most frequently implicated.[3]

The symptoms of the disease are a progressive loss of condition resulting in loss of bloom from the coat and in the skin becoming 'hide bound'. There is generally diarrhoea, but not invariably, and the appetite is usually retained until the animal becomes very weak. Mucous membranes will often appear anaemic. In very severe infestations deaths may occur suddenly.

The anaemia appears to be of a normochromic normocytic type, as there is a reduction in packed cell volume, haemoglobin content and red cell content, but no change in mean corpuscular volume or mean corpuscular haemoglobin concentration. There is a decline in the serum protein level, serum albumin decreasing markedly but serum globulin increasing. In the abomasal fluid, the concentration of sodium ions increases and the concentration of potassium and chloride ions decreases, resulting in an increased abomasal pH. Similarly, the plasma pepsinogen (the precursor of pepsin) levels are raised.[4]

Calves are most susceptible, but adult cattle, which are normally resistant, may develop the disease under conditions of poor nutrition, or under the strain of lactation and pregnancy, or if infestation is very severe.

The eggs, which are laid in large numbers by adult worms, pass to the ground with the faeces of the infected animal and hatch into larvae, which, after moulting twice, are capable of infecting another animal within 4 to 5 days, though this period may be prolonged in cold weather. These third stage larvae leave the faeces and migrate into the herbage where they await the grazing host. Most cattle harbour some of the parasites, but their health is only affected when a heavy infestation occurs.

The main requirements for larval development on the pasture are moisture, warmth and oxygen. Humid weather conditions and long herbage favour larval development, whereas drought and short herbage are less suitable. Nevertheless, a heavy infestation may occur in short dry pasture, as close grazing favours the ingestion of larvae from the lower parts of the herbage. Small paddocks frequently used for the grazing of calves near to the buildings are particularly dangerous.

There is no doubt that overstocking is the most important factor in influencing the development of an outbreak of the disease. A few larvae may survive for 12 months in a permanently moist and protected environment, such as deep in grass tufts, but most die within a few days. A period of 6 weeks should be sufficient before a heavily infested pasture is safe for susceptible animals, but where the herbage is long, and in damp cloudy weather, a rest of 5 to 6 months may be necessary.

Under conventional systems of rearing calves, it is suggested that resistance to the disease starts at the age of 4 to 6 months and develops to a maximum at 18 months to 2 years. However, certain immunological advantages may stem from rearing calves at pasture from birth. In 3 years of experiments[5] calves reared at pasture from birth showed an elevated rectal temperature from 10 to 25 days of age—this pyrexia began a few days after the time when pasture is usually first ingested by the young calf and it was not observed in calves kept indoors for the first 6 weeks of life and then put out to pasture. The faecal egg counts of trichostrongyles in calves on pasture from birth increased rapidly to 13 weeks of age, then at a very much slower rate until 21 weeks and thereafter declined, whereas the counts for calves put out to pasture at 6 weeks of age increased rapidly to very high values at 26 weeks of age. Moreover, in the latter treatment, the older the calf when it first became infected, the greater was the increase in egg counts with age. It seems possible, therefore, that the very small increase occurring in output of eggs after 13 weeks, for calves that had been exposed to parasites since birth, was associated with a development of immunity, for it is well known that the immune mechanism may restrict egg production of existing worms.[6] Similarly, a heavy larval challenge with Ostertagia to 6 month old calves that had been sucked at pasture and exposed to a natural field infection from birth failed to produce any lesions and only small numbers of larvae became established as mature worms.[7] The important point appears to be that calves should be managed so that they ingest a gradually increasing amount of grass from a young age.

Under poor nutritional conditions and when exposed to sudden very heavy infestation the resistance may break down. The autumn is the most dangerous time, as pastures may be heavily infested and the nutritional quality low. Even if heavily infected store animals are taken into buildings, their condition will gradually deteriorate if they are maintained on a low plane of nutrition. A recent study showed two types of clinical infection, one occurring in calves from late July to the end of the autumn grazing season, and the other being delayed for up to 6 months after grazing an infested pasture from late September onwards. Although while grazing there were no clinical signs in this second type of infection, the cattle were heavily infected with Ostertagia, which were inhibited in their early 4th larval stage in the gastric glands of the abomasal mucosa, and months later, maturation of the larvae caused clinical disease in both housed and outwintered stock.[4]

Supplementation of calves on pasture with a 16% crude protein concentrate mixture at the rate of 1 kg/day (2·2 lb/day) has prevented clinical signs in calves challenged with larvae, but the changes in the blood chemistry (see above) still occurred due, presumably, to a subclinical infection.[8]

Prevention of the condition can thus be afforded by feeding stock adequately, by avoiding overstocking and by the prophylactic use of phenothiazine. For prophylactic purposes, young cattle should receive treatment on one or two occasions during the summer and autumn and a final treatment before being housed in the autumn. The dose varies from 15 to 25 g depending on the age of the calf. In bright sunny weather, it is usually suggested that calves should be kept indoors for 48 hr after treatment, as otherwise damage to the eyes may be caused. After treatment the urine becomes pink in colour due to the presence of a red dye in the product.

For curative treatment during an outbreak of acute gastro-enteritis, two doses of phenothiazine should be given at intervals of a week, followed by one or two further doses at fortnightly intervals. Phenothiazine is less efficinet in cattle than in sheep and alternative drugs which are more efficient against some worms than phenothiazine have been introduced, in particular the organic phosphorus compounds and methyridine.

Of the organic phosphorus compounds, haloxan is effective orally against the mature and young adult Ostertagia worms, but is less effective against the larvae, especially against the mucosal 4th larval stage. Unlike other organo-phosphorus compounds, acute toxicity symptoms occur only with very high doses. It is recommended for routine use in young calves at grass in late summer and autumn and is given by drench.[9]

Methyridine can be given by mouth or by intraperitoneal or subcutaneous injection and is effective against most worms which cause parasitic gastro-enteritis. No other anthelmintic should be given within 24 hr of its use, and methyridine should not be used within 2 weeks following the use of an organic phosphorus compound.

When curative treatment is given, all animals in a group should be treated and after treatment should be transferred to a clean pasture.

PARASITIC BRONCHITIS (HUSK)

This condition is caused by a heavy infestation with a nematode worm, *Dictyocaulus viviparus*. The larvae of this worm enter the animal's body by way of the mouth from the pasture herbage; they lose their protective sheaths in the rumen, and then pass to the small intestine where, after penetrating the mucous membrane, they make their way into the lymphatic system and thence into the blood stream. The larvae are then carried by the blood stream to the lungs where they break through to the alveoli and migrate into the smaller bronchial tubes.

In this site, the larvae grow to maturity and produce eggs about 3 to 4 weeks after infection has taken place. Pulmonary lesions may be found seven days after infection,[10] i.e. 2 weeks before the worms become adult and give rise to larvae in the faeces. The eggs hatch into larvae which are coughed up, swallowed and then pass to the ground in the faeces. On first reaching the ground the larvae are not infective to other animals but, after passing through several larval stages, they become infective after a minimum period of 5 days.

Calves are most susceptible to the condition, and yearlings frequently become infected. It is probable that immunity to the condition develops in a similar way to that against parasitic gastro-enteritis (see p. 145). Calves sucking their dams at pasture seem less susceptible than bucket-fed calves, and it appears that Channel Island cattle are particularly susceptible to the acute form of the disease.

The main symptom of the complaint is a husky cough, which is slight and infrequent when the parasites are few in number. In chronic cases there is emaciation, loss of appetite, a staring coat, diarrhoea, pallor of the mucous membrane and sunken eyes. These symptoms may be manifested for as long as 4 to 5 months before death or recovery takes place. In acute cases, the cough is deeper and more husky, and fits of coughing occur, with great difficulty in breathing. The respiration and pulse rate are hurried, the head is extended, the tongue often protrudes, and mucus flows from the nostrils and corners of the mouth. The animal may even fall prostrate as a result of partial suffocation. Coughing fits may occur several times in a day, and single worms or masses of worms and streaks of blood may sometimes be seen in the mucous discharge.

Husk is associated with damp, low-lying permanent pastures. Nevertheless, the husk worm may also be present on well-drained dry pastures, especially where the herbage is thick and long. There is also considerable danger of infection from contaminated water supplies.

By ploughing up grassland, the husk worm larvae can be disposed of more effectively than the larvae responsible for gastro-enteritis, which tend to make their way upwards through the soil again. Spreading of dung pats by chain harrowing may be dangerous as, by improving the aeration of the dung, the young larvae in the centre of the faecal mass are helped in their development. The use of the chain harrow should therefore be avoided except in dry weather and when the herbage is short and the surface of the ground exposed.

Small fields and orchards adjacent to farm buildings are particularly dangerous places as they tend to be so frequently used by young animals that they become heavily infested. Manure heaps and dirty stockyards where a small amount of herbage is growing are also sources of infection. Manure heaps should be fenced and rainwater which has washed the surface of the heap should be removed by proper drainage. Infected manure should not be used on grassland.

Where a pasture is heavily infested, very considerable benefit will be obtained by grazing with non-susceptible animals such as horses and sheep. Pastures from which cattle have been completely excluded throughout the winter are usually free from infection in the spring, but this is not necessarily so, for in the climatic conditions of the west of Scotland lungworms may survive for 13 months.[11]

For the prevention of the condition, it is essential to maintain the cattle on a reasonable level of nutrition and protect them against a heavy infestation. In some localities, the risk of severe infestation among calves is so great that it is best to keep the animals in buildings during their first summer. Directly calves out at grass show symptoms of the disease, they should be taken off the infected pasture and housed. When indoors, the calves should be fed on a high plane of nutrition so that they have a good chance of overcoming the infection. Although constant moving of cattle from pasture to pasture is usually recommended as a help towards the limitation of infestation, research work[12] has indicated that, to build up an immunity, calves should be exposed to a low level of herbage infestation and then continue to have constant access to that infected herbage. Where risk of husk infection is great, concentrates should be fed to young stock as soon as the quality of the pasture starts to deteriorate.

Treatment with phenothiazine, similar to that recommended for parasitic gastro-enteritis, may be of benefit, as the two conditions are often present at the same time. By curing parasitic gastro-enteritis, the calf may be able to recover from the husk infection. Alternatively, injections into the trachea have been found to be effective in some cases.

More recently, these methods have been superseded by two drugs, which may be given orally or by injection. The first, cyanacethydrazide causes the worms to become inert and moved up the trachea by ciliary action.[13] However, it must be borne in mind that deaths of calves may sometimes occur as a result of lesions produced by the larvae during migration. The effect of the drug on immature worms appears to be limited. For subcutaneous injection, 15 mg/kg (0·7 g/100 lb) up to a maximum total dose of 5 g can be used, whereas by mouth 18 mg/kg body weight (0·8 g/100 lb) up to the same maximum can be given. A single treatment should be sufficient, but with animals severely infected, treatment should be carried out on three consecutive days. Animals should not be excited during treatment.

The second drug, diethylcarbamazine, is highly effective against immature worms, but almost without effect against adult worms. There is some difference of opinion as to whether treatment with this drug reduces immunity to subsequent infection. The drug was given at a level of 300 mg/kg live weight (13·6 g/100 lb) to 7½–16 week old calves 14 days after challenge with larvae; after a second challenge more worms were recovered from the treated calves than from the controls.[14] With a lower dose rate, however, no difference in immunity to a second infection was found.[15]

An oral vaccine is available as a protection against husk. This vaccine is based on husk worm larvae, irradiated by X-rays to make them sufficiently weak so that they do not migrate further than the mesenteric lymph nodes but sufficiently strong to stimulate immunity.[16] Two vaccinations are necessary. The first can be given after the calf is 6 to 8 weeks old and the second a month later. Immunity should be developed within 2 to 4 weeks after the second vaccination. However, even with vaccinated calves, resistance may break down if they are suddenly exposed to a heavy infection.

RINGWORM

This fungal disease in cattle is caused mainly by *Trichophyton verrucosum.* The very resistant spores become lodged in cracks on the surface of the skin and in the hair follicles. After germination the fungal hyphae grow on the inside or outside of the hair shaft, and the hair breaks off. The earliest lesion is a localised raised area on the skin on which the hair stands up. This gradually develops in size, becomes sharply defined and covered with grey or yellowish-grey scales which increase in thickness. The hair on the affected part then breaks off above the scaly crust and leaves a bare patch. The crusts in turn fall off naturally in 1 or 2 months, leaving a bald patch, which finally gives place to a new growth of hair. The main sites of the lesions are the head, neck and anal region, and they often give rise to pronounced itching.

Calves and yearlings are most susceptible and the disease may develop under conditions of poor nutrition, cold, wet, and worm or lice infestation. The complaint is more prevalent during the winter months and it is possible that bright sunlight may exert an inhibiting action on the growth of the fungus. Recovery from one ringworm infection usually confers immunity against a subsequent infection.

Where calves in a poor state of nutrition are affected, the disease may last for a year or more, but older animals usually make a speedy recovery when placed out of doors in bright, sunny weather and when they have access to spring grass.

Opinions differ as to the efficacy of various ringworm remedies which may be obtained from any veterinary chemist. By washing the cattle with a dilute solution of formalin (2%), containing soap, or a detergent that is not detrimental to human skin, the spread of infective spores to other parts of the body or to other individuals may be reduced. During this process the crusts may be removed and scraped into a container. It may be necessary to soften the crusts by applying a mixture of equal parts of soft soap and lard, and to remove the crusts on the following day. The lesions or healthy skin around them should not be scrubbed or rubbed hard, as this has been shown to increase the spread of infection. The ringworm ointment should then be applied and a second application made, if necessary, a week later. To prevent the spread of infection, treated animals should be segregated and the premises disinfected. Particular attention should be paid to parts of the buildings against which the infected animals may have rubbed. The premises should be scraped down, scrubbed with hot detergent or soda solution and later flamed with a blow lamp. Bedding should be burned and all woodwork creosoted.

On account of the transmissibility of the disease to man, overalls and gloves should be worn while dressing the animals and these should be disinfected after use. Similarly the hands, arms and neck should be washed and disinfected. Grooming equipment should be disinfected by immersion in 2% formalin for 24 hr.

Recent research has indicated the possibility of giving by mouth a metabolic product, called griseofulvin, of a certain species of penicillium, to prevent and cure ringworm.[17] This drug is incorporated into freshly formed skin and hair and inhibits the growth of the fungus in these tissues. The dose rate is about 33–44 mg/kg body weight (15–20 mg/lb) daily for a period of up to 3 weeks.

OX WARBLE FLIES

Two species of fly, *Hypoderma bovis* and *Hypoderma lineatum*, attack cattle in Great Britain. The adult insects are fairly large, dark, hairy flies with bands of yellow or orange which give them a superficial resemblance to humble bees. *H. lineatum* is on the wing earlier (May) than *H. bovis*, and is more prevalent in the west country. Like many flies, their activity is restricted during periods of rain and high winds, and moreover their cruising range appears to be limited to about 400 yards. As warble flies fly at a low altitude and appear to dislike crossing water, high hedges and rivers help to restrict the area of their activity.

The flies lay their eggs on the hair of the legs and bellies of cattle, those of *Hypoderma bovis* being laid singly and those of *H. lineatum* in rows of up to 20. After 4 or 5 days the eggs hatch into maggots, which penetrate the skin of the leg by means of their sharp mouth hooks and in the case of *H. lineatum* move up to the wall of the oesophagus where they spend the winter. In February, they leave the oesophagus and pass to the loins where they encyst until the pure maggot stage is reached. Some of the *H. bovis* maggots pass through the spinal canal on the way to the back. Once in the back each maggot makes a breathing hole. The swellings which result from the encysted maggots increase in size, soften and burst, discharging the maggot on to the ground where it pupates and becomes an adult after about 6 weeks (see Fig. 11.1).

The economic importance of the warble fly lies in the number of ways in which it causes damage. The flies themselves produce much excitement in stock, causing failure to put on flesh normally, and injury to the cattle from their efforts to evade attack. The hides are damaged in their most valuable part where the leather is thickest, and not infrequently large and painful abscesses are formed under the skin. In addition, losses are suffered by the meat industry as the migrating larvae leave greenish-yellow tracts containing a jelly-like material in the tissues, and these have to be trimmed from the carcase.

Although it has not yet been definitely ascertained whether a degree of resistance is developed by cattle, it seems that certain animals are able to escape the attention of the fly. Calves and young cattle are attacked more frequently and more severely than older beasts, which appear to have a degree of resistance, due possibly to their tougher hides.

Control at present depends upon destroying the larvae before they emerge from the skin of the back. Although the habits of the fly tend to keep it in a localised area and although benefit may accrue from treatment of cattle on any particular farm, eradication of the fly can only be achieved if treatment is done on every farm in the country.

Under the Warble Fly (Dressing of Cattle) Order, 1948 and the 1960 Amendment, it is compulsory to treat cattle visibly infested with the

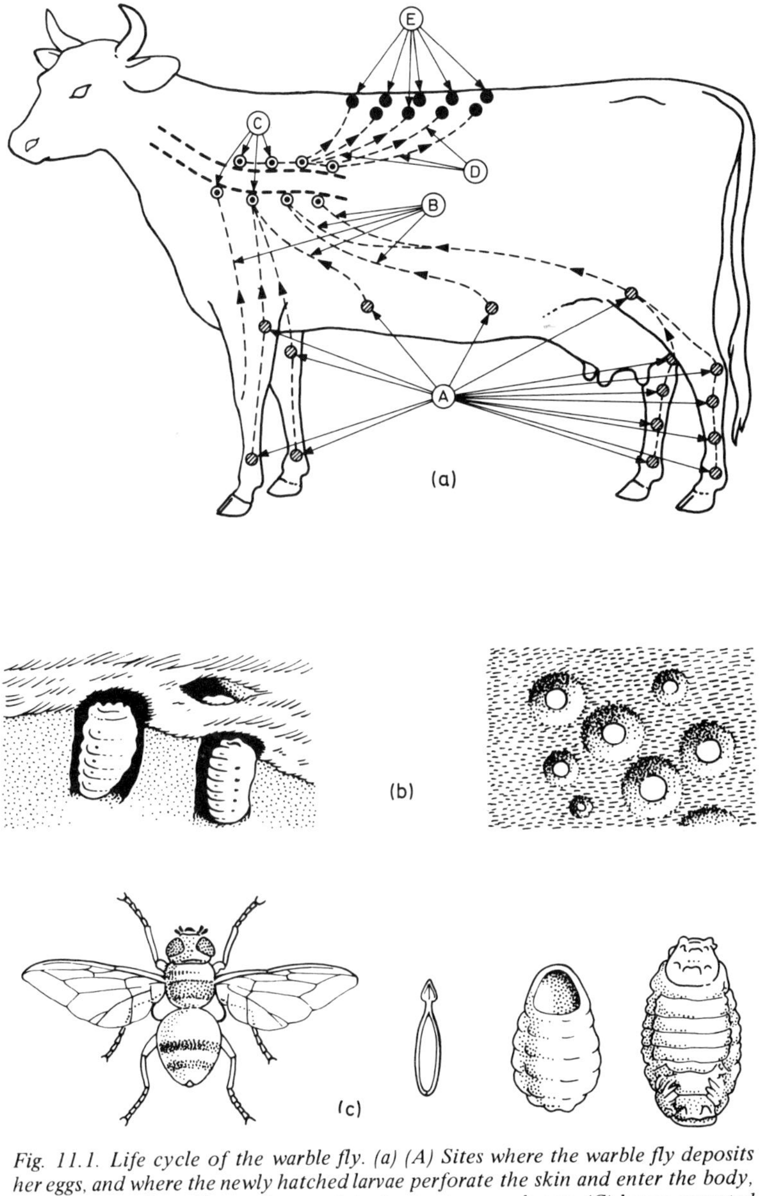

Fig. 11.1. Life cycle of the warble fly. (a) (A) Sites where the warble fly deposits her eggs, and where the newly hatched larvae perforate the skin and enter the body, (B) first migration of larvae from point of entry to oesophagus, (C) larvae encysted around the oesophagus from August to February, (D) and (E) second migration of larvae from oesophagus to back: February to June. (b) Warbled hide, section and plan views. (c) Ox warble fly with egg and larva

maggots of the warble fly and all cattle in which infection by the maggot can be ascertained by handling the animal. The Order states that all infected cattle must be treated as follows.

Treatment should start each year between 15th and 22nd March or as soon after as the larvae appear under the skin of the backs of the cattle, and should be repeated at intervals of not less than 27 days and not more than 32 days as long as larvae continue to appear, but not after 30th June. The dressing prescribed is 43 g (1½ oz) of derris resins or 14 g (½ oz) of rotenone and a sufficient quantity of dispersing agent to produce a stable emulsion or suspension and to ensure wetting the treated skin. If the derris wash is properly applied and the exudation or scab which forms on the warble holes is removed with a stiff brush to enable the derris to penetrate, every maggot should be killed.

However, the introduction of systemic organo-phosphorus insecticides[18–23] which control the maggots before they reach the back, should reduce the amount of dressing necessary under the Warble Fly Order. However, if warbles do appear on the back after using a systemic insecticide the cattle must be dressed externally. The organo-phosphorus compounds are suitable for beef cattle of all ages, but must not be administered within 60 days of slaughter. They may be given to dairy cows and heifers when dry, but may not be given during lactation or within 60 days of the beginning of lactation. Oral treatment should be given between the beginning of September and the end of November.

Metriphonate (Trichlorphon) may be given as a single dose by mouth or as a single dressing early in the spring. It is best to avoid treatment during December, January and February to prevent possible side effects as a result of destruction of large numbers of larvae while located along the spinal chord. Metriphonate may be given as a drench at a rate of 80 mg/kg (3·6 g/100 lb) body weight; this method, although killing a large proportion of migrating larvae, may cause slight toxic effects in a few animals. This drug is also effective against the nematode parasites causing parasitic gastro-enteritis in cattle.

REFERENCES

1. NEWMAN, A. J. *Vet. Rec.* **79**, 240 (1966)
2. FITZGERALD, P. R. *J. Protozool.* **12**, 215 (1965)
3. MICHEL, J. F. *Parasitology* **53**, 63 (1963)
4. ANDERSON, N., ARMOUR, J., JARRETT, W. F. H., JENNINGS, F. W., RITCHIE, J. S. D. and URQUHART, G. M. *Vet. Rec.* **77**, 1196 (1965)
5. ROY, J. H. B. *Vet. Rec.* **76**, 511 (1964)
6. TAYLOR, E. L. *Outl. Agric.* **3**, 139 (1961)
7. ROSS, J. G. and DOW, C. *Br. vet. J.* **120**, 279 (1964)
8. LELAND, S. E. Jr., DRUDGE, J. H. and DILLARD, R. P. *Am. J. vet. Res.* **27**, 1555 (1966)

9. ARMOUR, J. *Vet. Rec.* **76**, 1364 (1964)
10. JARRETT, W. F. H., McINTYRE, W. I. M. and URQUHART, G. M. *J. Path. Bact.* **73**, 183 (1957)
11. JARRETT, W. F. H., McINTYRE, W. I. M. and URQUHART, G. M. *Vet. Rec.* **66**, 665 (1954)
12. MICHEL, J. F. *Vet. Rec.* **69**, 1118 (1957)
13. WALLEY, J. K. *Vet. Rec.* **69**, 815, 850 (1957)
14. KENDALL, S. B. *J. comp. Path.* **75**, 443 (1965)
15. CORNWALL, R. L. *Res. vet. Sci.* **4**, 435 (1963)
16. JARRETT, W. F. H., McINTYRE, W. I. M., JENNINGS, F. W. and MULLIGAN, W. *Vet. Rec.* **69**, 1329 (1957)
17. LAUDER, I. M. and O'SULLIVAN, J. G. *Vet. Rec.* **70**, 949 (1958)
18. HARRISON, I. R. *Vet. Rec.* **70**, 849 (1958)
19. KENNY, J. E. and THORNBERRY, H. *Irish vet. J.* **12**, 198 (1958)
20. TURNER, E. C. JR. and GAINES, S. A. *J. econ. Ent.* **51**, 582 (1958)
21. Veterinary Investigation Service and Central Veterinary Laboratory, Ministry of Agriculture, Fisheries and Food *Vet. Rec.* **74**, 127 (1962)
22. BROWN, F. G. *Vet. Rec.* **74**, 577 (1962)
23. BROWN, F. G. *Vet. Rec.* **75**, 585 (1963)

Index